The Structure and Dynamics of Geographic I

90 0933047 X

Oxford Series in Ecology and Evolution

Edited by Paul H. Harvey and Robert M. May

The Structure and Dynamics of Geographic Ranges

KEVIN J. GASTON
University of Sheffield

OXFORD
UNIVERSITY PRESS

OXFORD

UNIVERSITY PRESS

Great Clarendon Street, Oxford OX2 6DP

Oxford University Press is a department of the University of Oxford.
It furthers the University's objective of excellence in research, scholarship,
and education by publishing worldwide in

Oxford New York

Auckland Bangkok Buenos Aires Cape Town Chennai
Dar es Salaam Delhi Hong Kong Istanbul Karachi Kolkata
Kuala Lumpur Madrid Melbourne Mexico City Mumbai Nairobi
São Paulo Shanghai Taipei Tokyo Toronto

Oxford is a registered trade mark of Oxford University Press
in the UK and in certain other countries

Published in the United States
by Oxford University Press Inc., New York

Database right Oxford University Press (maker)

First published 2003

A catalogue record for this title is available
from the British Library

Library of Congress Cataloging in Publication Data
(Data available).

ISBN 0 19 852640 7 (hardback)
ISBN 0 19 852641 5 (paperback)

10 9 8 7 6 5 4 3 2 1

Typeset by Cepha Imaging Pvt Ltd.
Printed in Great Britain
on acid-free paper by T. J. International Ltd., Padstow

To Sian

Preface

Probably all of the academics that I know have at least one of what are, euphemistically, known as 'on-going' projects. Most have several. These are not those projects on which they labour every day, nor indeed every week or month, but those which, if they are nurtured, grow principally by the slow and more measured acquisition of material—an odd fact here, a handy quote there, a paradigm-changing graphic somewhere, and a sprinkling of potentially interesting but as yet unread references. The infant mortality of such projects is typically very high, most seldom progressing much beyond the first fits of enthusiasm. For the others there are brief growth spurts, but more commonly periods of virtual dormancy, if not ossification. As is the way of such things, most of these survivors are still destined never to reach maturity, but to be laid to rest in the bin of might-have-beens, the sack of nearly-becames, and on the wayside of retirements. Some, whilst never themselves reaching much beyond adolescence, contribute to others that did attain adulthood, donating insights without acclaim or even recognition. A few, however, eventually attain majority in their own right, and are finally exposed to public gaze.

This book has been one of my on-going projects for a number of years. Not that it even vaguely resembled a book for most of that period. Rather, it comprised a steadily expanding collection of reprints, copies of papers, book chapters, bits of theses, and an eclectic set of typed, scrawled, or virtually illegible notes that bore close testimony to the remarkable diversity of suppliers of stationery to my various employers. Occasionally it was exposed to the light, in a fit of enthusiastic organization, or a desperate search for a half-forgotten thought or observation. More usually, it occupied a variety of box files and filing trays. It has travelled widely, from office to office, and institution to institution, but seldom without some concerns for its welfare and a desire to find it a satisfactory new resting place.

The opportunity to transform this material into a book was provided by The Royal Society, London, which has generously supported my research over several years under a fellowship scheme that is marvellous in its simplicity and in the academic freedom that it provides. Only thus have I been able to find the time really to come to grips with my, sometimes hard won, treasures.

Books are seldom written in isolation, and this one has certainly not been so. First, I am grateful to a number of people who have, over the years, encouraged my pursuit of an understanding of the structure of geographic ranges. Foremost amongst these is John Lawton, who originally sparked my curiosity about the nature of the distributions of species, and then took opportunities to fan the flames or helped place me in the way of other people who would do likewise. Others include Tim Blackburn, Steven Chown, Bob May, John Spicer, and Mark Williamson, all of whom have had a profound influence on my thinking about the world of ecology. They have also been wonderful collaborators. Tim Blackburn, in particular, deserves special mention; he has been all that one could ask of a collaborator and coauthor.

Tim Blackburn, Steven Chown, David Coltman, Paul Harvey, Fangliang He, Sian Gaston, Sarah Jackson, Alex Jones, Hefin Jones, Owen Petchey, and David Storch generously undertook to read some or all of the manuscript, providing comments that were variously challenging, insightful, and pedantic, but were always offered in a spirit of encouragement. Yet others have been kind enough to suffer their brains to be picked, to partake of my need to bounce ideas around, or have helped me track down, at times particularly obscure, pieces of information. In particular, I would thank Andy Brewer, Andy Clarke, Nigel Collar, Chris Humphries, Melodie McGeoch, Allan Mee, John Reynolds, Ana Rodrigues, Michael Usher, Dick Vane-Wright, and Tom Webb. Parts of this book were written during journeys to Gough Island and Chile. I am grateful to Steven Chown, Miriam Fernandez, Niek Gremmen, Pablo Marquet, and Peter Ryan for providing stimulating and enjoyable company during those trips.

Paul Harvey and Bob May were generous in their encouragement of this project and in their thoughtful advice, and Ian Sherman oversaw its publication with his usual mix of energy and cajoling.

As always, I am grateful to Sian and Megan for their seemingly endless founts of enthusiasm and support.

Sheffield   K. J. G.
January 2002

Contents

1

Introduction

I have lately been especially attending to Geograph. Distrib., & most splendid sport it is,—a grand game of chess with the world for a Board.

C. Darwin (1856–58)

1.1 The case of the green-backed heron

The first green-backed heron *Butorides striatus* (L.) that I ever saw was standing motionless on a steaming hot day, amidst a clump of mangroves growing at a remote site on the Pacific coast of Costa Rica. In so doing, it gave decidedly better views than many other birds that I encountered on the journey that provided my introduction to the tropics. Since then I have observed this small, compact-looking, dark species of heron at innumerable sites around the world, on humid days in the Gambia and cool ones on the eastern coast of Australia, skulking around a small lake in Argentina, and flying boldly across a major river in South Africa.

The sites at which I have seen the green-backed heron have been characterized principally by their geographic dispersion. Indeed, the spatial distribution of this species extends across parts of the Nearctic, Neotropic, Afrotropic (including Madagascar), Indotropic, and the Australasian biogeographic regions, in a broad latitudinal band from about 50°N to about 40°S (Fig. 1.1; Cramp 1977; Harrison *et al.* 1997a). In this sense, the green-backed heron appears to be a highly unusual bird (or heron for that matter); most others are much more restricted in their occurrence, typically being confined to only one of these regions. But, just what are the general patterns of variation in the geographic range sizes of species, and how does the green-backed heron fit?

Individuals of the green-backed heron can be found in a wide variety of aquatic habitats, both fresh and saltwater. These include rivers, estuaries, lagoons, streams, lakes, ponds, swamps, marshes, mangroves, mudflats, harbours, intertidal zones, and open flood plains (Cramp 1977; del Hoyo *et al.* 1992; Harrison *et al.* 1997a). This embraces much, albeit not quite all, of the breadth of habitats used by herons more generally, and there must therefore be some constraint on the distribution of the green-backed heron simply because it is a heron. However, many of these habitats occur widely outside the geographic distribution of this species, so what determines the limits of its occurrence?

Beyond its already wide geographic distribution, vagrant individuals of the green-backed heron have also made it to such far-flung places as Bermuda,

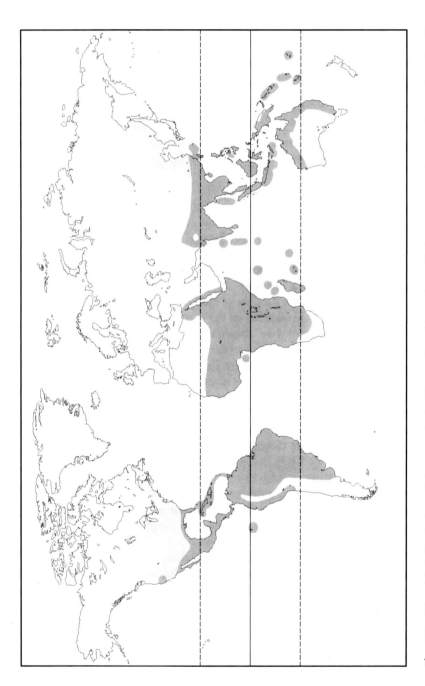

Fig. 1.1 The global distribution of the green-backed heron *Butorides striatus*. Dark grey—area where species tends to be all year round, and light grey—area habitually occupied for breeding but where species is not normally present outside breeding season. From del Hoyo *et al.* (1992).

Manitoba, Nova Scotia, Newfoundland, Greenland, the Azores, Britain, and way out into the Pacific (Cramp 1977; Lewington *et al.* 1991; del Hoyo *et al.* 1992). Indeed, although mainly sedentary, most northerly races are migratory, in some regions the species is regarded as something of a wanderer, and post-breeding dispersal has been known to lead to population eruptions. So what role does dispersal play in the geographic range of the green-backed heron?

Within its geographic range the green-backed heron has been recorded across a broad band of elevations. Whilst mostly found in the lowlands (including along sea coasts), it occurs at up to 1500 m in Madagascar, regularly up to 3000 m in Colombia and Ecuador, and even up to 4050 m in Peru (del Hoyo *et al.* 1992). To what degree is the ability of the species to exploit this elevational range associated with the broad latitudinal and longitudinal extent of its geographic distribution?

The green-backed heron exhibits extensive variation in plumage, in body size, in its ecology and in its reproductive biology; it is also one of the few species of birds known to fish with the use of bait, a behaviour that has been recorded in several parts of its range (Higuchi 1986; del Hoyo *et al.* 1992). Thirty or more subspecies of *B. striatus* are commonly recognized (del Hoyo *et al.* 1992). Of these, two are often raised to the status of full species in their own right, *B. s. virescens* in North America (as far south as Panama) and *B. s. sundevalli* in the Galapagos (e.g. Sibley and Monroe 1990; Monroe and Sibley 1993); the latter has a very restricted breeding range (Stattersfield *et al.* 1998). This results in two, still reasonably widely distributed species being distinguished, the striated heron *B. striatus* (L.) and the green heron *B. virescens* (L.). However, there is often far from unanimity in such matters, and these proposals are controversial. The status of other subspecies remains equally question-able. Regardless, the existence of such genetic variation as to result in the recognition of so many subspecies begs the question as to why there is not sufficient variation to enable the species (or species-complex) to spread its bounds yet more widely. Why hasn't the green-backed heron continued to evolve to the point that it has overcome existing barriers to its geographic distribution and become yet more widespread?

Because of its skulking habits, the green-backed heron is a difficult species to census, but it is considered globally to be common and not to be particularly threatened with large scale extinction. Its range also appears not to have changed markedly in recent times. Whilst there is rather little comparable information, it is plain that the local abun-dance of the green-backed heron varies markedly across its geographic range. Thus, in the United States, its breeding density is highest in the extreme south-east (especially Florida; Price *et al.* 1995), while in southern Africa it is most frequently reported from the Okavango (Harrison *et al.* 1997a). Does this variation reflect some broader pattern, and what determines the abundance structure of the range of this species?

1.2 Areography

All of the above questions have been asked about the geographic range of the green-backed heron. They, and many related questions, might equally be asked about any other of the estimated 13 million or so extant species of organism, albeit most are not so widespread and their distributions and details of their ecology are far more poorly known. Indeed, they could also be asked about any of the still much greater number of species that have, at some point in the history of life, become globally extinct (extant species comprise perhaps 2 per cent of all the species that have ever lived). That is, the questions can be framed as very general ones about the ways in which species, and the individuals of which they are comprised, are distributed at geographic scales and what might be the determinants of these patterns.

To date, there have been a large number of case studies that have attempted to answer one or more of these questions for particular species. That is, they have addressed what I shall term the structure of geographic ranges. These include studies of the form and determinants of the positions of range edges, of the extent and shape of spatial distributions, and of the pattern of abundance within the limits to occurrence. Thus, for example, there have been numerous studies of the ecologies of individual species explicitly at the margins of their geographic ranges (Table 1.1). These have concerned representatives of groups as diverse as trees, grasses, corals, molluscs, flies, beetles, fish, toads, birds, and mammals, a far broader range than have been the subjects for the investigation of many other issues in ecology. The diversity of issues addressed is also impressive, including the factors that might limit the geographic distributions of species, the nature of population dynamics at range limits, patterns of growth in body size, and patterns of genetic variation at the margins. Indeed, these studies have discovered such, at first glance, apparently disparate observations as that Indian crested porcupines *Hystrix indica* do not occur further north because of insufficient nocturnal foraging time (Alkon and Saltz 1988), whilst painted turtles *Chrysemys picta* do not do so because low temperatures kill the hatchlings (St Clair and Gregory 1990), that white-spotted charr *Salvelinus leucomaenis* may be deformed in populations at the edge of its distribution (Morita and Yamamoto 2000), that individuals of the small-leaved lime *Tilia cordata* do not reproduce for many kilometres towards its range limit (Pigott 1989), and that genetic variability in peripheral populations of the sand scorpion *Paruroctonus mesaensis* is low (Yamashita and Polis 1995).

Interest in such, sometimes surprising, matters derives from many quarters. Studies of species at their geographic range limits have, for example, been performed in a variety of contexts, including those of forestry, fisheries, and pest control, as well as being driven by sheer curiosity. This interest has also persisted for a very long time. Questions of how and why individual species are distributed in the ways that they are follow from basic observations of the natural world and have come to many minds over the ages. There are doubtless many thousands of published papers and books that impinge on them.

Significantly, however, in recent decades the emphasis has been shifting from what historically has predominantly (though far from exclusively) been the steady accumulation of case-by-case accounts for particular species to the determination of broader generalizations about the structure of geographic ranges. This has come to be termed the field of *areography*, defined succinctly by Rapoport (1982, p.1)

Table 1.1 Studies of features of the ecology and evolutionary biology of particular species at, usually some part of, their geographic range limits

Species	Feature	Study
Pteridophytes and Gymnosperms		
Killarney fern *Trichomanes speciosum*	Population structure	Rumsey *et al.* (1999)
Jack pine *Pinus banksiana*	Growth, reproduction, age structure, and fire history	Desponts and Payette (1992), Houle and Filion (1993), Conkey *et al.* (1995), Despland and Houle (1997)
Lodgepole pine *Pinus contorta*	Genetic variation	Yeh and Layton (1979)
Ponderosa pine *Pinus ponderosa*	Genetic structure	Hamrick *et al.* (1989)
Red pine *Pinus resinosa*	Age structure and fire history	Bergeron and Gagnon (1987), Bergeron and Brisson (1990)
Pitch pine *Pinus rigida*	Age structure, fire dynamics and phytosociology	Meilleur *et al.* (1997)
Scots pine *Pinus sylvestris*	Regeneration, recruitment, and mortality	Kullman (1992), Stöcklin and Körner (1999)
Black spruce *Picea mariana*	Environmental constraints and growth	Bonan and Sirois (1992), Payette and Delwaide (1994)
Red spruce *Picea rubens*	Population structure and climatic responses	Webb *et al.* (1993)
Eastern hemlock *Tsuga canadensis*	Age structure, growth rate, and site preference	Kavanagh and Kellman (1986)
Angiosperms		
Douglas-fir dwarf mistletoe *Arceuthobium douglasii*	Environmental constraints	Smith (1972)
Gambel oak *Quercus gambelii*	Environmental constraints	Neilson and Wullstein (1983)
Small-leafed lime *Tilia cordata*	Distribution, history, seed sterility and age	Pigott and Huntley (1978, 1980, 1981), Pigott (1981, 1989)
Arizona willow *Salix arizonica*	Herbivory	Maschiniski (2001)
Hackberry *Celtis occidentalis*	Population structure and growth	Houle and Bouchard (1990)
Barrel cactus *Ferocactus acanthodes*	Habitat	Ehleringer and House (1984)

(continued)

Table 1.1 *(continued)*

Species	Feature	Study
[buttercup] *Hepatica acutiloba*	Pollen limitation	Murphy and Vasseur (1995)
Fastigiate gypsophila *Gypsophila fastigiata*	Population dynamics	Bengtsson (2000)
Common fumana *Fumana procumbens*	Population dynamics	Bengtsson (1993)
Strapwort *Corrigiola litoralis*	Genetic variation	Durka (1999)
Nottingham catchfly *Silene nutans*	Genetic variation	van Rossum *et al.* (1997)
[gentian] *Gentianella germanica*	Reproduction	Luijten *et al.* (1998)
Prickly lettuce *Lactuca serriola*	Abundance	Prince *et al.* (1985)
[grass] *Corynephorus canescens*	Factors limiting survival	Marshall (1968)
[sedges] *Carex stans* and *C. ensifolia*	Genet age	Jónsdóttir *et al.* (2000)
Annual phlox *Phlox drummondii*	Population dynamics	Levin and Clay (1984)
Lloydia *Lloydia serotina*	Reproduction and genetic variation	Jones *et al.* (2001)
Invertebrates		
[coral] *Pocillopora damicornis*	Associated fauna	Black and Prince (1983)
[barnacle] *Balanus balanoides*	Limitation by temperature	Barnes (1957, 1958)
[barnacle] *Chthalamus stellatus*	Distribution	Southward and Crisp (1954)
[bivalve] *Elliptio complanata*	Genetic structure	Kat (1982)
[bivalve] *Lampsilis radiata*	Genetic structure	Kat (1982)
[bivalve] *Abra tenuis*	Growth and population dynamics	Dekker and Beukema (1993)
[bivalve] *Pisidium amnicum*	Growth and population dynamics	Araujo *et al.* (1999)
Baltic clam *Macoma balthica*	Respiration	Hummel *et al.* (2000)
Roman snail *Helix pomara*	Genetic structure	Järvinen *et al.* (1976)
Sand scorpion *Paruroctonus mesaensis*	Genetic structure	Yamashita and Polis (1995)
[spider] *Atypus affinis*	Genetic structure	Pedersen and Loeschcke (2001)
[tick] *Ixodes ricinus*	Impact of climate change	Lindgren *et al.* (2000)
[bush cricket] *Tettigonia canrans*	Distribution	Panelius (1978)
Thimbleberry aphid *Masonaphis maxima*	Population dynamics	Gilbert (1980)

(continued)

Table 1.1 *(continued)*

Species	Feature	Study
[halictid bee] *Augochlorella striata*	Social structure	Packer (1990)
[skipper butterfly] *Parnara guttata*	Life table	Matsumura (1992)
[drosophilid fly] *Drosophila subobscura*	Genetic structure	Kohonencorish *et al.* (1985)
[drosophilid fly] *Drosophila serrata*	Genetic structure, cold response and phenotypic variation	Blows and Hoffmann (1993), Blows (1993), Jenkins and Hoffmann (1999, 2000)
European corn borer *Pyrausta nubilalis*	Population characteristics	Chiang (1961)
Southern pine beetle *Dendroctonus frontalis*	Limitation by temperature	Ungerer *et al.* (1999)
Fish		
Bluespotted sunfish *Enneacanthus gloriosus*	Habitat and life history	Peterson and van der Kooy (1997), Snyder and Peterson (1999)
Rainbow trout *Oncorhynchus mykiss*	Abundance and distribution	Flebbe (1994)
Brown trout *Salmo trutta*	Abundance and distribution	Flebbe (1994)
Brook trout *Salvelinus fontinalis*	Abundance and distribution, effects of climate change	Meisner (1990), Flebbe (1993, 1994)
White-spotted charr *Salvelinus leucomaenis*	Limitation by temperature morphological deformities	Nakano *et al.* (1996), Morita and Yamamoto (2000)
Dolly varden *Salvelinus malma*	Limitation by temperature	Nakano *et al.* (1996)
[sculpin] *Cottus nozawae*	Distribution	Yagami and Goto (2000)
Sea snail *Liparis liparis*	Population stability	Henderson and Seaby (1999)
Amphibians and reptiles		
[tree frog] *Hyla arborea*	Predation	Brönmark and Edenhamn (1994)
Natterjack toad *Bufo calamita*	Phylogeography	Rowe *et al.* (1998)
American toad *Bufo americanus*	Activity	Bider and Morrison (1981)

(continued)

Table 1.1 *(continued)*

Species	Feature	Study
Fowler's toad *Bufo woodhousii fowleri*	Age structure	Kellner and Green (1995)
Montandon's newt *Triturus montandoni*	Morphometrics	Dandova *et al.* (1998)
Painted turtle *Chrysemys picta*	Environmental constraints	St Clair and Gregory (1990)
Blanding's turtle *Emydoidea blandingii*	Behaviour and genetic variation	Standing *et al.* (1997), Mockford *et al.* (1999)
Collared lizard *Crotaphytus collaris*	Size and growth	Sexton *et al.* (1992)
Birds		
Gentoo penguin *Pygoscelis papua*	Reproduction	Bost and Jouventin (1991)
Ringed plover *Charadrius hiaticula*	Nest predation	Pienkowski (1984)
Barn owl *Tyto alba*	Reproduction	Marti (1997)
Green woodpecker *Picus viridis*	Habitat selection	Rolstad *et al.* (2000)
Pied flycatcher *Ficedula hypoleuca*	Reproduction	Järvinen and Väisänen (1984)
Goldcrest *Regulus regulus*	Incubation energetics	Haftorn (1978)
Hume's leaf warbler *Phylloscopus humei*	Behaviour and ecology	Gross and Price (2000)
Wood warbler *Phylloscopus sibilatrix*	Limitation by temperature	Tiainen *et al.* (1983)
Willow warbler *Phylloscopus trochilus*	Limitation by temperature	Tiainen *et al.* (1983)
Chiffchaff *Phylloscopus collybita*	Limitation by temperature	Tiainen *et al.* (1983)
Robin *Erithacus rubecula*	Abundance, morphology, and condition	Pérez-Tris *et al.* (2000)
Great tit *Parus major*	Breeding biology	Veistola *et al.* (1995)
Cardinal *Richmondena cardinalis*	Home range and habitat	Dow (1969)
Mammals		
Nine-banded armadillo *Dasypus novemcinctus*	Limiting factors	Taulman and Robbins (1996)
American bison *Bison bison*	Limitation by snow	Daubenmire (1985)
Indian crested porcupine *Hystrix indica*	Limitation by foraging time	Alkon and Saltz (1988)

(continued)

Table 1.1   *(continued)*

Species	Feature	Study
Yellow-bellied glider *Petaurus australis*	Feeding behaviour	Carthew *et al.* (1999)
Franklin's ground squirrel *Spermophilus franklinii*	Distribution	Johnson and Choromankinorris (1992)
Eurasian badger *Meles meles*	Environmental constraints	Kauhala (1995), Virgós and Casanovas (1999)
Red fox *Vulpes vulpes*	Limitation by resources and competition	Hersteinsson and Macdonald (1992)
Arctic fox *Alopex lagopus*	Limitation by resources and competition	Hersteinsson and Macdonald (1992)

as 'the study of the geographical range of taxa'. However, this definition is perhaps a little too all encompassing and it might better be framed as the study of the structure of the geographic ranges of taxa. Certainly this seems to fit with Rapoport's own approach to the subject. The seeds of this search for generalizations were sown by some of the influential figures of ecology and evolutionary biology of the past, including C. Elton, G. E. Hutchinson, E. Mayr, R. H. MacArthur, and C. B. Williams.

The present book presents an overview of present understanding of the structure of geographic ranges, and hence it is about some of the fundamental problems that areography sets out to resolve. Areography itself can perhaps be seen as a component of the recently emerged subject of macroecology, the study of the division of food and space among species at large spatial and temporal scales (Brown and Maurer 1989; Brown 1995; Maurer 1999; Gaston and Blackburn 2000).

The current heightened interest in macroecology, and more particularly in making generalizations about the structure of geographic ranges (areography), reflects two things, necessity and opportunity.

1.2.1 Necessity

Some of the major issues attracting attention in ecology at the present time concern phenomena at large spatial scales (distinguishing between cartographic and colloquial definitions of scale, because small scale according to the former is large scale according to the latter (Curran *et al.* 1997), here and elsewhere scale is used in the colloquial sense as a synonym of words such as size and area; small scale refers to a small area, and large scale to a large area). These include the need to predict and respond to the effects of global environmental change (Peters and Darling 1985; Peters and Lovejoy 1992; Kareiva *et al.* 1993; Lawton 2000), to maintain biodiversity (Wilson and Peter 1988; Gaston 1996a; Reaka-Kudla *et al.* 1997; Gaston and Spicer 1998), to control diseases and their vectors (Randolph and Rogers 2000; Rogers and Randolph 2000), and to understand and thwart the spread of alien invasive species (Williamson 1996; Ruiz *et al.* 1997; Pimentel *et al.* 2000; Blackburn and Duncan 2001a, 2001b).

All of these areas have become of increasing concern with growth in the global human population, and the associated growth in resource demands, in habitat destruction, and in the movement of people and goods. That is, concern has grown with growth in the 'human enterprise' (Ehrlich 1995; Gaston and Spicer 1998). The statistics are stark, the human population numbers six billion and continues to increase, it uses, co-opts or destroys approximately 40 per cent of all potential terrestrial net primary productivity (Vitousek *et al.* 1986), has come to dominate the vast majority of the area of most biomes (Hannah *et al.* 1995), has elevated atmospheric carbon dioxide levels by more than 25 per cent above pre-industrial levels (Carbon Dioxide Information Analysis Centre 2000), and has introduced perhaps 400 000 species into areas in which naturally they did not occur (Pimentel 2001). Thus, recent studies have sought, for example, to determine how malaria might spread under global climate change (Rogers and Randolph 2000), the effectiveness of 'flagship' species in identifying areas that are important for the conservation of species in general (Williams *et al.* 2000), and the relationship between human densities and the occurrence of various carnivore species (Woodroffe 2000).

Alongside this attention has been a growing realization that, more generally, many other issues in population and community ecology cannot satisfactorily be understood without both a small scale and a broad, or macroecological, perspective (Levin 1992; Brown 1995; Maurer 1999; Gaston and Blackburn 2000; Lawton 2000). This realization appears to have come about amongst workers studying terrestrial, freshwater, and marine systems alike. Studies at small scales fail to reveal the influences that arise from the regional context in which an assemblage is embedded, and which determines such things as the pool of species available for colonization, and places constraints on the likelihood of species and individuals with particular traits arriving and establishing.

Unfortunately, a focus on large spatial scales is at odds with that which historically has been, and which continues to the present to be, predominant in ecology (Fig. 1.2; Maurer 1999; Gaston and Blackburn 2000; Lawton 2000). Rather, ecological studies have focused on small areas. In particular, the subject has been dominated by experiments, and, since Darwin (1888) established his 'worm stone' to determine the rates at which earthworms bury material, the scale of these experiments has typically been extremely small. Even large-scale ecological experiments are small scale by most other standards, although there is an encouraging trend, at least, toward experiments that are repeated in multiple areas distributed across a region (e.g. Hector *et al.* 1999).

By way of example of the difficulties that result from an emphasis on small scales in ecology, Root and Schneider (1993) observe the mismatch between the poor resolution of global climate models (typically 500×500 km, and difficult to refine because of the number of computations required) and the fine resolution of study plots

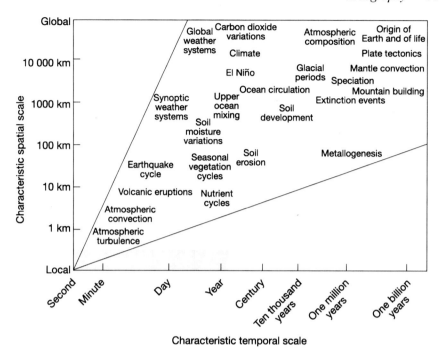

Fig. 1.2 Characteristic temporal and spatial scales of Earth system processes. The vast majority of ecological studies fall within the shaded area. Modified from National Aeronautics and Space Administration (1988).

in ecological field work (typically at most 'tennis-court sized'). Costs and logistics notwithstanding, they argue strongly that ecological studies over broad geographic and long temporal scales are needed to bridge this gap, if the responses of species to climatic change are to be adequately and reliably predicted (see also Pace 1993).

Such calls for attention to be paid to large-scale ecological phenomena inevitably places issues of the structure of geographic ranges in the spotlight. Thus, questions about how species are likely to respond to global environmental change beg an understanding of what determines the limits to their geographic ranges (e.g. Davis *et al.* 1998a; 1998b; Thomas *et al.* 1998), identifying the best course for maintaining biodiversity suggests a need to know how widely species are distributed and what influences this extent (e.g. Gaston 1994a; IUCN 1994), the need to battle against diseases of humans, livestock, and crops requires an understanding of why species are more abundant in some places than they are in others (e.g. Rogers 1979; Rogers and Randolph 1986; Rogers and Williams 1994; Randolph and Rogers 2000), and the spread of alien invasive species raises issues of how best to predict the patterns and rates of expansion of geographic ranges (Hengeveld 1989; Williamson 1996; Chown *et al.* 1998).

Most of these questions are not new. Indeed, many were posed long ago (e.g. Elton 1927; Williams 1964; MacArthur 1972). However, under the twin banners of areography and macroecology they are being revisited and examined more intensely and in much greater detail.

1.2.2 Opportunity

As well as the growing necessity of developing an understanding of the structure of geographic ranges, the opportunities to do so have also improved markedly within the past decade or so, and are continuing, making this perhaps the most exciting time to be undertaking such work. Of course, it is difficult to disentangle cause and effect in this growth, the extent to which necessity has proven the motherhood of invention and to which opportunities have seen exploitation. Nonetheless, it is readily apparent that it has become easier to obtain the requisite data for analysing problems of geographic range structure, and the tools with which to analyse this information have developed apace.

The spatial scale of the study of geographic ranges renders it very difficult, although not necessarily impossible, for the individual researcher or a small team to obtain first-hand the data needed to answer many of the most important questions. The time, effort, and resources required are often simply too great. Thus, it has taken the establishment and operation of a number of broad-scale sampling programmes, frequently involving large numbers of people (often including many amateurs) and with some centralized system for handling the resultant information, to yield the required data. Understanding of how best to operate such schemes has developed quickly (e.g. Smith 1990; Gibbons *et al.* 1993; Asher *et al.* 2001; Fox *et al.* 2001).

Information on the occurrences and abundances of species at multiple, disparate sites across one or more regions have resulted from a variety of increasingly sophisticated mapping and census programmes (Fig. 1.3). Some highlights of these, with an unashamed avian and United Kingdom bias (reflecting, in part, the fact that these are some of the best and the earliest established such programmes), include:

1. *International and national census or monitoring schemes*—the Christmas Bird Census (CBC) and the Breeding Bird Survey (BBS) in North America (Robbins *et al.* 1989; Sauer *et al.* 1997), the Common Birds Census (CBC) and the Breeding Bird Survey (BBS) in Britain (Marchant *et al.* 1990; Gregory *et al.* 1997), several constant effort bird mist-netting monitoring programmes (e.g. DeSante *et al.* 1995; Haapala and Saurola 1995; Peach *et al.* 1998), and the Rothamsted Insect Survey (Taylor 1986). In Britain alone, monitoring schemes are now operating for aphids, moths, butterflies, birds, and mammals.

2. *International and national atlas schemes*—for birds across southern Africa (Penry 1994; Harrison *et al.* 1997a, 1997b; Parker 1999), plants, amphibians, and reptiles, birds, and mammals across Europe (Jalas and Suominen 1972–94, Jalas *et al.* 1996; Gasc *et al.* 1997; Hagemeijer and Blair 1997; Mitchell-Jones *et al.* 1999), for birds across Australia (Blakers *et al.* 1984), and for vascular plants,

molluscs, grasshoppers, dragonflies, butterflies, amphibians and reptiles, birds, and mammals across Britain (Perring and Walters 1962; Marshall and Haes 1990; Arnold 1993, 1995; Gibbons *et al.* 1993; Merritt *et al.* 1996; Kerney 1999; Asher *et al.* 2001).[1]

3. *State, provincial and county atlas schemes*—for birds in regions of Britain (e.g. Harding 1979; Smith *et al.* 1993; Dennis 1996; Standley *et al.* 1996; R.D. Murray *et al.* 1998), of Canada and the United States (e.g. Cadman *et al.* 1987; Brewer *et al.* 1991; Branning 1993; Hess *et al.* 2000), of South Africa (Cyrus and Robson 1980), and of Australia (Saunders and Ingram 1995); a detailed listing of North American avian atlases can be found at http://www.americanbirding.org/norac/atlaspubld.htm.

4. *City atlas schemes*—for butterflies in Sheffield, London, and Manchester (e.g. Garland 1982; Plant 1988; Hardy 1998).

The mapping schemes have revealed many fascinating features of the distributions of species, including commonalities of patterns of occurrence, disjunctions, abrupt limits, patchiness, unexplained absences, and so on. Likewise, the census schemes have revealed previously unsuspected patterns in local population dynamics, of spatial variation in abundances, and of covariance in population fluctuations.

Biological information of possible importance in interpreting abundance and distribution data has also increasingly become available, in the form of collations of body size, life-history, and status information for particular groups of organisms (e.g. Dunning 1984, 1992; Damuth 1987; Grime *et al.* 1988; Collar *et al.* 1994; Silva and Downing 1995; Stotz *et al.* 1996; see also Peters 1983). Thus, for example,

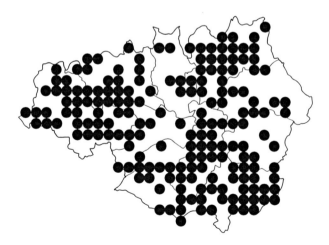

Fig. 1.3(a) The distribution of the small skipper *Thymelicus sylvestris* butterfly across the city of Manchester (United Kingdom). From Hardy (1998).

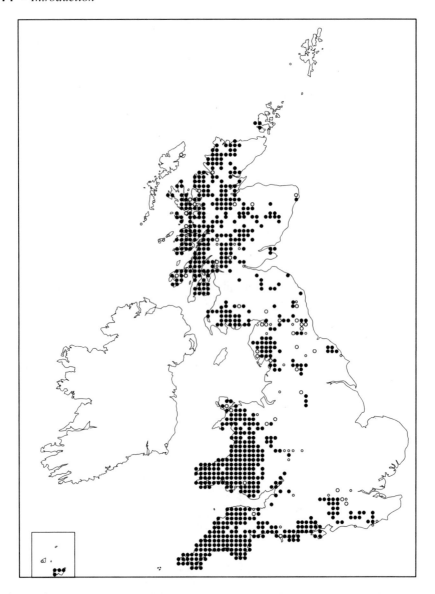

Fig. 1.3(b) The distribution of the golden-ringed dragonfly *Cordulegaster boltonii* across Britain and Ireland (small open circles are records for pre-1950, large open circles for 1950–74, closed circles for 1975–90). From Merritt *et al.* (1996).

collations now exist of the body masses of a high proportion of the extant species of birds (Dunning 1992) and mammals (Silva and Downing 1995).

Likewise, the quality and accessibility of broad-scale environmental data has improved immeasurably, particularly as a result of the employment of remote

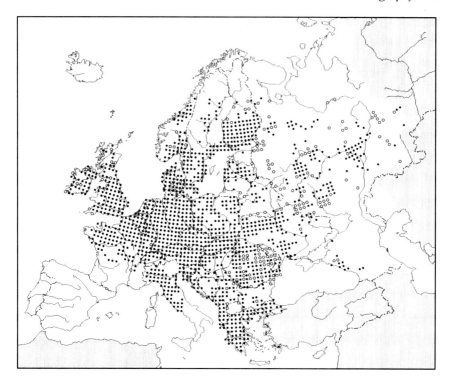

Fig. 1.3(c) The distribution of the smooth newt *Triturus vulgaris* across Europe (open circles are for records pre-1970, closed circles for post-1970, and grey areas were not mapped). From Gasc *et al.* (1997).

sensing techniques. Data are increasingly available on geology, terrain, soils, climate, productivity, and vegetation, and temporal changes in many such variables, for much, if not all, of the Earth's surface.

In the face of such dramatic improvements in the quality and availability of data appropriate for analysing and understanding the structure of geographic ranges it would, of course, be wrong to give the impression that this has resolved the information requirements in this area. First, most of this information derives from a few wealthy, developed nations, whose biota is very unrepresentative of that of the world at large. Second, heavy reliance has still to be placed on less adequate sources. Indeed, much of the work seeking to determine generalities in range structure has required the compilation of information from a diverse variety of pre-existing sources. This carries with it some significant constraints. In outlining the methods taken in her studies of the latitudinal spans of seaweed species, Pielou (1977a, p.301) put the case well, writing that 'As in nearly all biogeographic investigations, the data come ready-made from the literature; inevitably, much had to be taken on trust' and that '… if biogeographic studies are to proceed at all, one must assume that

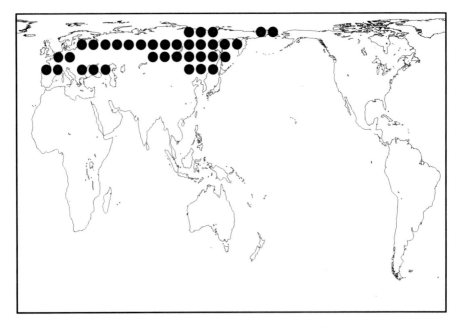

Fig. 1.3(d) The distribution of the bumblebee *Bombus sichelii* across the world. From Williams (1998).

the signal-to-noise ratio of the data is high enough to assure that, by appropriate statistical analysis, the signal may be recovered and correct generalizations derived.'

Even for data sets that have been obtained in a more systematic fashion, for example from directed sampling schemes, problems of replication, bias, and accuracy persist. Thus, for example, despite the increasingly high quality of many distributional atlases, and a growing number of areas for which species of the same taxon have been mapped more than once (e.g. breeding birds in Britain and France; Sharrock 1976; Yeatman 1976; Gibbons *et al.* 1993; Yeatman-Bertholet and Jarry 1994), cases remain scarce where the methodology has been replicated in such a way as to make genuine changes in patterns of occurrence unequivocally distinguishable from differences in sampling effort.

The improved availability of powerful computing facilities has given some assistance in overcoming these constraints, particularly by enabling more detailed exploration of possible biases in data sets arising from non-biological sources. More generally, such facilities have provided the means to manipulate and analyse the data sets that are commonly associated with areographical studies, and particularly to use spatially explicit analytical methods (e.g. those embodied in Geographic Information Systems; Johnston 1998). The size of some data sets is incredibly large. CONABIO, the national biodiversity body for Mexico has computerized and georeferenced data on more than 4 250 000 specimens of plants and animals (Soberón and

Koleff 1997). The Biological Record Centre of the Centre for Ecology and Hydrology, United Kingdom, holds in digital form over 6 000 000 records for 15 000 species (Anon 1999). Likewise, in the preparation of the most recent atlas of breeding birds in Britain and Ireland, for the years 1988–1991, a total of 551 370 records were submitted for the presence of a species in a 10×10 km square; 275 732 of the records were non-duplicate (i.e. a single record of a species in a particular square; Gibbons *et al.* 1993). Even the most recent atlas of dragonflies for this same region and resolution, albeit for a far longer period, employed 108 801 records (Merritt *et al.* 1996).

Spatially explicit methods of data analysis have yet to have a truly marked impact on areographical studies, and yet there is little doubt that they will do so. A whole range of tools are now available, principally developed in the field of geostatistics, which enable data on the regional abundances and distributions of species to be examined with much greater rigour than has hitherto commonly been the case (e.g. Isaaks and Srivastava 1989; Borcard *et al.* 1992; Maurer 1994; Carroll and Pearson 2000; Lennon 2000). Combined with advances in mathematical and simulation modelling they herald a new era in the study of the geographic ranges of species.

1.3 This book

Responding in part to this stimulating mix of necessity and opportunity, and building on the questions about the distribution of the green-backed heron asked at the outset of this chapter, this book provides an overview of the structure of geographic ranges of species, as it is understood at present. It is comprised of four subsequent chapters. The first three of these each address a particular facet of this structure. These facets have been chosen because they seem to me to embrace the major issues of concern in studying ranges, albeit not necessarily as always recognized by those working in individual fields of investigation. Chapter 2 concerns range edges, and in particular what determines their spatial position, and Chapter 3 the size of geographic ranges, the extent to which this varies, the frequencies of different sizes, and patterns in the variation. A more complex aspect of the structure of ranges is considered in Chapter 4, namely the abundance structure, including the frequency of local abundances of different magnitude, intraspecific abundance–range size relationships, changes in local abundance along environmental gradients, and the spatial relations of local abundances of different magnitude. The final chapter (Chapter 5) draws out some of the principal, applied implications from the material in the preceding chapters.

Put simply, the goals of this book are five-fold:

1. *To establish the basic patterns in the structure of geographic ranges and their inter-relationships.* A number of patterns have been suggested (Table 1.2), although the full breadth of available evidence has seldom been brought to bear on them. Drawing together those patterns that withstand close scrutiny highlights a number of connections between them that have not, or have only poorly, been drawn in the past.

Table 1.2 Patterns that have been suggested to occur in the structure of the geographic ranges of species, and the sections of the book in which those patterns are discussed

Pattern	Section
Intraspecific range edge–genetic variation relationship	2.4
Interspecific species-range size distribution	3.4
Interspecific range size–speciation relationship	3.4.1
Interspecific range size–time relationship	3.4.2
Interspecific range size–extinction relationships	3.4.3
Interspecific range size–latitude relationship	3.5.4
Interspecific range size–longitude relationship	3.5.5
Interspecific range size–trophic group	3.5.6
Interspecific range size–body size relationship	3.5.7
Interspecific range size–dispersal ability relationship	3.5.8
Interspecific abundance–range size relationship	3.5.9
Interspecific range size–niche breadth relationship	3.5.9
Interspecific range size–niche position relationship	3.5.9
Interspecific range size–genetic variation relationship	3.5.10
Intraspecific mean–variance (density) relationship	4.2
Intraspecific abundance distribution	4.2
Intraspecific abundance–range size relationship	4.3
Intraspecific occupancy–environment relationship	4.4
Intraspecific abundance–environment relationship	4.4
Intraspecific abundance surface	4.5

2. *To synthesize information on the structure of geographic ranges from different fields.* The need for such a synthesis extends as far as simply drawing together botanical and zoological examples and those with pure and applied roots, around which have developed different literatures, with different emphases and different underlying paradigms. Some of the best data on range structure have arisen from studies in agroecosytems, forestry, and fisheries, and yet this work is seldom, if ever, cited in the broader ecological or biogeographic literatures.

3. *To provide ready access to examples of different facets of the structure of geographic ranges.* Published studies are scattered in the literature to a remarkable degree, frequently some of the best can be found in rather unexpected places and are not identifiable as being particularly relevant based on titles and key words. Indeed, much of the literature on range structure does not emerge from electronic data bases and search engines, even using a wide array of apparently appropriate terms. I would venture to predict, for example, that most of those working on general patterns in the structure of geographic ranges will have previously encountered only a small to moderate proportion of the case studies of range limits listed in Table 1.1. The bibliography of this book is therefore an important component. To improve readability, limits have been imposed on the numbers of

studies cited in support of particular statements. To maintain more complete coverage of the literature, particularly sources of useful historical or empirical examples, additional studies are listed in the Notes section at the end of the book, with their presence being indicated in the text by superscripted numbers.

4. *To illustrate that understanding of the structure of geographic ranges is substantially more advanced than many seem to assume.* Repeatedly, the literature makes simplistic and unsupported assertions about the structure of geographic ranges. Here, a more considered approach is attempted.

5. *To highlight the implications of the structure of geographic ranges.* As already mentioned, part of the motivation behind many studies of range structure has been concerns of an applied nature. Those implications will be explicitly explored in their own right, rather than solely as adjuncts to other issues.

Throughout, attention is focused on the entire geographic ranges of species (see Section 3.3 for further discussion of the components of geographic ranges), rather than some small parts thereof. Inevitably, however, material is drawn from a host of studies, whose perspectives were often much narrower (indeed, some provide areographical insights which the original authors chose not to develop or of which they were perhaps not aware). In the context of macroecological studies in particular, a distinction is drawn between 'partial' analyses, which are performed over areas which embrace the entire geographic ranges of none or only a small proportion of the species concerned, and 'comprehensive' analyses, which are performed over areas which embrace all or a very large proportion of the extents of the geographic ranges of the species concerned (Gaston and Blackburn 1996a). What is sought here is an understanding of the latter, although this may in some cases require consideration of the former.

2

Range edges

The tendency of animals and plants to multiply beyond the means of subsistence and to spread over all available areas is well understood. What naturalists wish to know is not how species are dispersed, but how they are checked in their efforts to over-run the earth.

C. H. Merriam (1894, p. 229)

2.1 Introduction

Merriam (1894) clearly regarded the question of what determines the limits to the geographic ranges of species to be more pertinent to naturalists (for which we might here read 'ecologists') than that of what enables them to spread. Twenty years later, Griggs (1914, p. 44) stated that with regard to the Sugar Grove region of Ohio, 'For many of the plants which terminate here, especially those which I know best because their boundaries cross the area, it is quite beyond my power of analysis to discover any reason why they should find their limits where they do and not go a few miles further'. Another 75 years on, Grubb (1989, p. 15) could still write, albeit more generally, 'Given the complexities, it is perhaps not surprising that so few exact studies have been made on the determination of distribution-limits, but, until more work ... is done, it will not be practicable to formulate anything more than the vaguest generalisations in this field'. But, just what is the present state of understanding?

A huge variety of factors have been argued to limit the spatial distributions of particular species, including rivers, temperature, snow cover, resource availability, natural enemies, unstable population dynamics, and insufficient genetic variability. To make some sense of this profusion, various classifications have been made of the ways in which the geographic ranges of species might be constrained. These include distinctions between limits posed by abiotic and by biotic factors, between constraints posed by unmodifiable environmental factors, modifiable environmental factors, and natural enemies, or between limits that result from physiological or ecological adaptation and those that result from extinction/colonization phenomena (for discussion of these and other schemes see MacArthur 1972; Emlen *et al.* 1986; Caughley *et al.* 1988; Brown and Lomolino 1998; Hochberg and Ives 1999).

Such schemes are of some heuristic value. It is, however, perhaps more useful initially to recognize that the question of what limits the geographic ranges of species, or what determines the positions of their borders, can be addressed at different levels.

First, one can ask whether there are abiotic and/or biotic factors that prevent further spread, and what these might be. These factors might include physical barriers (e.g. mountain chains, rivers), climatic factors, the absence of essential resources, or the impact of competitors, predators, or parasites.

Second, one can consider what limits the geographic range of a species in terms of how its local population dynamics change at the edge of its geographic range such that it is unable to persist beyond this point. It will persist in an area (though not necessarily successfully reproduce there) all the while the net addition of individuals exceeds the net loss. At range edges abiotic and/or biotic factors must be acting to ensure that this inequality is not met. But, this may be for a variety of population dynamic reasons.

Third, one can address the genetic mechanisms that prevent a species from becoming more widespread and expanding the limits to its geographic range. Abiotic and biotic factors are only limiting because a species has not evolved the morphological, physiological, or ecological capacities to overcome them. The question is why this should be so.

Range limits will not be understood without considering all three of these levels.

In this chapter, these three different levels at which the determinants of the limits of the geographic ranges of species can be addressed will each be considered in turn. In sympathy with the general heterogeneity of terminology used in studies of range edges, I will treat as synonymous reference to range borders, boundaries, edges, limits, margins, and peripheries. All of these terms will be used interchangeably. However, two broad kinds of range edges can be distinguished. There are those that are internal to the range, defining areas within the overall extent of occurrence of the species from which it is absent (e.g. altitudinal and habitat limits). Then there are those that are external, in as much as they define the outermost limits to this extent of occurrence. Some have regarded both as constituting 'marginal' populations (e.g. Carson 1959; Antonovics 1976). In this chapter, and throughout this book, I shall predominantly be concerned with external limits, although reference will be made to occasional examples concerning internal ones (particularly those associated with altitude). The degree to which general lessons about the determinants of internal limits can be extrapolated to external is largely unknown (but see Carter and Prince 1985a, b; Shreeve *et al.* 1996). It seems inescapable that the same principles will apply, but very doubtful that in any given case one could necessarily extrapolate the details, particularly because internal limits may tend to be associated with steeper environmental gradients than external ones.

Throughout what follows, two things must be borne in mind. One is that the spatial structure of geographic range limits is typically complex. The other is that the spatial position of range limits moves through time. Several schemes have been suggested for understanding the structure of range limits (e.g. Salisbury 1926; Chiang 1961; Antonovics 1976; Brussard 1984; Emlen *et al.* 1986; Gorodkov 1986). In one such scheme, Gorodkov (1986) distinguishes a number of different kinds of limits to the geographic range of a species (Fig. 2.1), in order of increasing distance

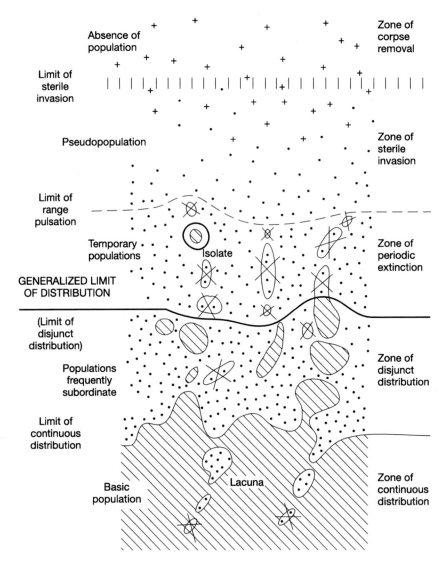

Fig. 2.1 A schematic presentation of the structure of the limits to geographic ranges. Areas with pluses indicate moribund individuals; dots, living individuals; crosses, periodically invading populations; and oblique lining, stable or constant populations. See text for further details. From Gorodkov (1986).

from the range centre:

1. *limit of zone of continuous distribution*—encircles the part of the range where the average regional climate is adequate for the species to maintain itself;

2. *limit of zone of disjunct distribution*—encircles the part of the range where the species can exist only in localities where the regional climate deviates from the mean;

3. *limit of zone of periodic extinction*—encircles the part of the range where, because of active or passive dispersal in favourable seasons or years, populations are produced that persist for a single or a few generations;

4. *limit of zone of sterile invasion*—encircles the part of the range where there is not even the possibility of a complete life cycle under favourable conditions and the death of immigrant individuals or their immediate progeny is inevitable;

5. *limit of zone of removal of remains of organisms*—encircles the region to which wind, water currents, and the like, may remove the remains of moribund or dying individuals (whilst not generally regarded as comprising part of the geographic range of a species, in some cases, such as in many paleontological studies, it may be difficult or impossible to recognize this region for what it is).

This scheme is, as any other would be, highly idealized, it centres on climatic reasons for limitation, ignores the fact that in no region are distributions actually continuous, and has difficulty in dealing with phenomena such as migration. But, the point remains that *the* edge of a geographic range does not exist. More troubling, it is not always obvious with which kind of edge individual published studies are primarily concerned.

Essentially, range limits are abstractions from reality. A wide variety of methods have been developed to define the edge of the geographic range of a species for analytical purposes. At one extreme lies the rather opaque, but probably the commonest, approach whereby an investigator simply draws a boundary based either on a map of known occurrences of a species, or some combination of known and presumed occurrence. Extreme outliers to the bulk of the ranges of species are not unusual, and their exclusion or inclusion will disproportionately influence where the position of the edge of a range is determined to lie (e.g. Smith 1972; IUCN 1994; Santelmann 1991; van Rossum *et al.* 1997). The variable status of populations (as exemplified in Gorodkov's scheme above) would doubtless also have an influence, if this were actually known, but it seldom is.

Simple rules have been employed in some cases to provide a more objective approach to drawing such boundaries. Thus, Curnutt *et al.* (1996) identified the range limit of a species by first distinguishing the cells of a 1° latitude×1° longitude grid in which at least one individual had been recorded at a survey site in the last 10 years, and then connecting all neighbouring outside grid squares either vertically, horizontally, or diagonally, incorporating gaps of one square in the edge vertically or horizontally but excluding gaps of more than one square.

At the other extreme to methods of defining the edges of geographic ranges lie more formal mathematical techniques, which are usually very transparent. Thus, Root (1988a, see also Root 1988b) plots the range limit of a species as lying at the point at which interpolated local abundance values attain 0.5 per cent of the maximum abundance value. Likewise, Price *et al.* (1995) identified the limit as the point at which interpolated local abundance values attain, depending on species, 0.5, 0.25 or 0.1 individuals per survey route per year (see also Mehlman 1997). Linder *et al.* (2000) use ordinary kriging (a procedure that interpolates values across a regular grid from irregularly spaced data points using information on the spatial autocovariation of these values) to generate abundance surfaces for North American birds, and then define the range boundary as the contour line of zero abundance. In exploring changes in the local abundances of species of birds towards their range limits in Britain, Blackburn *et al.* (1999) employed five measures of the position of sites relative to these limits, their northing, easting, average nearest neighbour distance (on the assumption that occurrences at range limits are more distant from neighbours), average distance to all unoccupied local areas, and average distance to all occupied local areas.

Although this is usually ignored, the position of the edge of a geographic range will depend on the spatial resolution at which that limit is mapped. Because maps of low spatial resolution are frequently used to plot range boundaries, there is a marked degree of uncertainty in their precise position. This may, for example, exaggerate the climatic range over which a species occurs by including sites, such as perhaps cold mountain peaks or hot valleys, outside the observed climate ranges of the species (Shao and Halpin 1995). One solution to the problem of spatial resolution is to think about range edges in terms of fractals (Maurer 1994). Other ways of identifying boundaries are emerging from other areas of spatial analysis (e.g. Jacquez *et al.* 2000).

Formal mathematical approaches to defining range edges have the advantage of being explicit in their formulation, but the disadvantage that typically they tend to be very demanding of data, and thus impossible to apply to the great majority of species.

The potentially detailed structure of range edges is additionally complicated by the fact that these limits are temporally dynamic (Sections 2.2.2, 3.4.2), whatever impressions to the contrary might be conveyed by standard maps of the occurrences of species (which tend to give no or little indication of the temporal frequency with which different areas are occupied, both because the data are usually lacking and because such dynamics are hard to represent simply). From season to season and year to year, the range edges of many species will move back and forth, latitudinally, longitudinally, and with elevation or depth. Over longer periods species will colonize fresh areas, and vacate others, commonly with systematic directionality in these movements.

Both the spatial complexity of range limits, and their temporal dynamism are widely ignored in considerations of what determines the position of these limits. However, this will not generate the complete understanding that is desired. Moreover, as will be seen, information on both of these issues may inform understanding of these determinants.

2.2 Abiotic and biotic factors

2.2.1 Physical barriers

Ostensibly, the simplest barriers to expansion of the geographic distributions of species, at least the simplest to comprehend, are physical or topographic. Indeed, many considerations of the determinants of range edges are concerned solely with species that are not subject to such constraints, apparently in the belief that physical barriers are readily understood.

Merriam (1894) considered physical barriers to the distributions of species to be rare. The exceptions were for terrestrial species in the case of oceans, and presumably for marine species in the case of land masses. The distributions of terrestrial species must ultimately be limited by the occurrence of rivers, lakes, and oceans, the distributions of freshwater species by intervening land and ocean, and the distributions of marine species by land. Thus, for example, the geographic ranges of many bird and primate species in the lowland forest regions of equatorial Africa and South America are limited by rivers in the large systems found in these areas, with sister species commonly being represented on opposite banks, particularly of large and fast-flowing rivers (Wallace 1849; Capparella 1991; Ayres and Clutton-Brock 1992; Eeley 1994; Eeley and Lawes 1999). This highlights the potential interaction between those forces that limit the geographic ranges of species, and those that are involved in the process of speciation itself (Section 3.4.1). Here, it seems probable that the riverine barriers that prevent species from expanding their ranges are the very same barriers that subdivided the geographic ranges of ancestral species to produce the daughters whose distributions are observed today (although given the temporal dynamism in the position of many Amazonian rivers, presumably the precise location of these barriers will have shifted through time). Depending on the model of speciation favoured, this may be a common phenomenon.

Mountain chains are known to have a profound effect on the distributions of lowland terrestrial species. Indeed, their significance as physical barriers has been argued to explain some important patterns of species diversity in the northern hemisphere, on the basis that they served to hinder the reinvasion of some regions post-glaciation. In a related vein, they also shape the orientation of ranges in some regions and groups. In temperate northern areas, there are very extensive land masses with rather small areas of mountains; a pine-forest-dwelling species can potentially have a range from Scotland to Cambodia. The major tropical areas are widely separated by intervening oceans so circumtropical spread is far less likely than a northern, circumpolar distribution. Many of the world's mountains are also in the tropics, further disrupting potential range sizes there.

In marine systems, ocean flows may effectively fulfil a similar role to some physical barriers in terrestrial ones. As Gaylord and Gaines (2000) observe, the fact that where major nearshore ocean currents collide they are typically composed of waters with quite different properties (e.g. temperature) has meant that the occurrence of species limits at these points has commonly been interpreted as a result of these differences. However, they can readily be explained in terms of the potential of ocean currents to limit geographic ranges, particularly of species with pelagic dispersing larvae.

Commonly the presence of physical barriers is reflected in the co-occurrence of the range limits of numbers of species (Rapoport 1982). However, such co-occurrence, which may define sharp discontinuities between biogeographic provinces, can result for a variety of other reasons and is not of itself indicative of such barriers (e.g. Anderson and Marcus 1992; Dahl 1992; Boone and Krohn 2000; Gaylord and Gaines 2000).

Barriers are commonly only such because the dispersal abilities of species are insufficient to overcome them. Thus, the influence of barriers and dispersal abilities on range limitation are tightly linked, albeit not synonymous. The more substantial barriers plainly act as such to some groups of species no matter how good their relative dispersal abilities. Dispersal limitation is likely to be more important at lesser barriers for those species with poor dispersal abilities (particularly those whose individuals are incapable of moving long distances under their own powers or via another agency), but it may also be important for those species with geographic ranges that are expanding either as a result of changes in environmental conditions or some phenotypic or genotypic change (i.e. they have yet to attain the geographic limits imposed by some other force). Limitations to dispersal may explain why the northward movement in the northern range limits of some species appears to lag behind the pace of change in climatic conditions, particularly where these exploit fragmented habitats (e.g. Gear and Huntley 1991; Payette 1993; Hill *et al.* 1999).

There have been a few explicit tests to distinguish between dispersal limitation and environmental restriction (e.g. Pierson and Mack 1990; Eriksson 1998). Thus, Pierson and Mack (1990) experimentally introduced populations of the alien cheatgrass *Bromus tectorum*, into mature forest stands in which it appears to attain a range limit. They found that its natural absence from these stands can largely be attributed to its intolerance of the environment rather than a lack of opportunity to reach such sites. Conversely, Eriksson (1998) concluded that the regional distribution of the dwarf-shrub *Thymus serpyllum* is likely to be limited by dispersal, with recruitment after seed sowing being equally good at sites with or without established populations.

It is apparent that dispersal abilities can limit the colonization of suitable sites even at a local scale (e.g. Lawton and Woodroffe 1991; Primack and Miao 1992; Thomas *et al.* 1992; Thomas 1994, 2000; Menéndez and Thomas 2000). If this is so, then dispersal limitation over yet greater distances seems likely to be very common. Brown (1987) documents one good example. In the 1930s, several individuals of the kangaroo rat *Dipodomys ordii* from Oklahoma were introduced, in an intentional experiment, to a sand dune on the shore of Lake Erie in Ohio, approximately 1200 km east of their range. A substantial population established and persisted until at least the early 1950s.

There is growing evidence that where previous barriers to dispersal are naturally overcome and the geographic ranges of species expand, the initial colonists have particularly well developed dispersal abilities. Thus, for example, the spread of the long-winged conehead *Conocephalus discolor*, a small, green bush-cricket in Britain, which is thought to be associated with ameliorating climate, has initially been by very long-winged and more mobile individuals capable of extended flights (Burton 2001).

Of course, populations of many species have been accidentally or intentionally successfully introduced to areas far removed from their native geographic ranges, underlining the significance of dispersal limitation. Indeed, the sheer scale of such movements is difficult to comprehend. In some developed countries, it seems likely that the movement of species by human agency has become as important, if not more so, than movement by natural means (e.g. Hodkinson and Thompson 1997). One consequence is increasingly to complicate understanding of the determinants of the natural distributions of species.

The apparent simplicity of the above observations about the role of physical barriers in range limitation hides many complexities of interpretation. Indeed, physical barriers exemplify the multiple levels of causality of range edges. First, they may themselves be seen as resulting directly in range limitation. Second, physical barriers may be seen only as being such because of the constraints that exist to the environmental tolerances or the dispersal abilities of species. If these were lessened, then neither the drastic changes in conditions represented by physical barriers (e.g. land to sea, lowlands to highlands), nor the difficulties of crossing them, might pose an obstacle to the distribution of a species. Third, they may be seen not simply as determinants of range edges, but perhaps as effectively causal in generating the species concerned in the first place, through allopatric speciation (Section 3.4.1).

2.2.2 Climate

Climate has long been held to play a significant role in limiting the geographic distributions of species (e.g. Merriam 1894; Hutchinson 1918)[1]. Indeed, arguably more has been written about climate as a limiting factor than any other. Two pieces of evidence are commonly cited in support of the belief in its importance. First, the limits to the distributions of species are often found to be coincident with particular combinations of climatic conditions. Second, the distributions of species are often seen to change through time in broad synchrony with changes in environmental conditions.

Coincidence with climatic conditions

The observation that the boundaries to the geographic ranges of species tend to coincide with the occurrence of identifiable sets of climatic conditions has repeatedly been made (e.g. Salisbury 1926; Good 1931; Jeffree 1959; Pigott 1975; Rogers and Randolph 1986; Caughley *et al.* 1987)[2]. Thus, for example, in a classic paper, Iversen (1944) observed that holly *Ilex aquifolium* is only present when mean winter temperature exceeds about –1°C (Fig. 2.2).

More recent papers examining relationships between range limits and climate continue a rich tradition. Thus, Root (1988b) investigated the association between environmental factors and distributional boundaries for 148 species of wintering land birds across North America. The comparisons, based on determining the average deviation between the position of a range edge and a contour for an environmental variable, revealed frequent correspondence between range limits and environmental factors (Fig. 2.3). For example, average minimum January temperature, mean length of frost-free period, and potential vegetation are frequently associated with northern range limits. Less than 1 per cent of all the associations observed were expected to occur by chance.

Jeffree and Jeffree (1994; see also Jeffree and Jeffree 1996) have shown that plots of summer against winter temperatures occurring at sites within a species' range tend to have a positive slope, and fall within a region with an approximately elliptical outline. They conclude on this basis that 'far from being limited by isothermal lines of summer or winter temperatures, the combination of temperatures *x* and *y* that constitutes the limits for a species changes continuously around the perimeter of the elliptical scatter' (there is also some suggestion of this in Fig. 2.2). This ignores the fact that a species does not occupy all the sites within its range, which could modify the shape of the ellipse, but the fundamental point remains that simple interpretations of the coincidence of climatic conditions and the occurrences of species may frequently be inadequate.

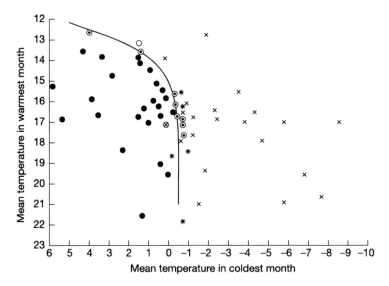

Fig. 2.2 Occurrence of holly *Ilex aquifolium* at meteorological stations with respect to mean temperature for the coldest month and for the warmest month (°C). Filled circles, holly occurs within the station area; open circles, holly sterile within the station area; cross, holly absent within the station area; circle within circle, station on boundary area of occurrence of holly; cross within circle, holly strayed into woods from gardens; and star, boundary area of occurrence of holly lies immediately outside station area. From Iversen (1944).

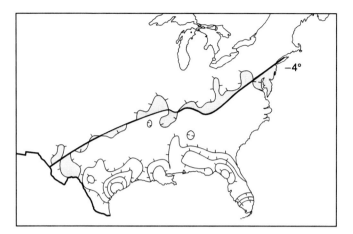

Fig. 2.3 Winter distribution and abundance of the Eastern phoebe *Sayornis phoebe* in North America, and the position of the –4°C average minimum January temperature isotherm. The contour intervals are 20, 40, 60, and 80 per cent of the maximum abundance value of 1.38 individuals seen per party hour (tick marks point down slope). The area between the isotherm and the northern range limit is indicated by the shading. From Root (1988b).

More broadly, it is now common place to develop models of the spatial occurrence (presence/absence or density) of a species based, at least in part, on spatial variation in climate (e.g. Hocker 1956; Beerling 1993; Huntley 1994; Sykes *et al.* 1996; Hill *et al.* 1999; Gioia and Pigott 2000)[3]. Thus, for example, Robinson *et al.* (1997a, 1997b) predicted the probable distribution of tsetse flies in unsurveyed areas by determining the environmental characteristics of areas of tsetse *Glossina* spp. presence and absence in surveyed areas. With respect to single environmental variables, the best predictions for *Glossina morsitans centralis* were made using the average normalized difference vegetation index (75 per cent correct predictions) and the average of the maximum temperature (70 per cent correct); for *G. m. morsitans* and *G. pallidipes* the best predictions were given by the maximum of the minimum temperature (84 per cent and 86 per cent correct, respectively) (Robinson *et al.* 1997a). The predictive power of models improved from the simple to the more complex, and the use of a principal component analysis of multivariate climate and remotely sensed vegetation data followed by a maximum likelihood classification resulted in predictions for *G. m. centralis* and *G. m. morsitans* of 92.8 per cent and 85.1 per cent, respectively (Robinson *et al.* 1997b).

Models, in the broadest sense, of the relationships between the occurrences of species and climatic and/or other environmental variables have been used to predict how far invading species (particularly pests) might spread (e.g. Cook 1924; Spradbery and Maywald 1992; Pheloung and Scott 1996; Pheloung *et al.* 1996; Higgins *et al.* 1999; Weber 2001)[4], the distributions of rare or restricted species (e.g. Jarvis and Robertson 1999), spatial patterns of species richness (e.g. Gioia and Pigott 2000), priority areas for conservation (Araújo and Williams 2000; Williams and Araújo 2000), the areas into which species (often of conservation or agricultural concern) might successfully be intentionally introduced (e.g. Booth *et al.* 1988; Samways *et al.* 1999; De La Ville *et al.* 1998), areas which species might successfully recolonize (e.g. Mladenoff *et al.* 1999), and how geographic distributions might respond to climate change (Section 5.4; e.g. Beerling 1993; Huntley 1994; Carey 1996; Jeffree and Jeffree 1996; Nakano *et al.* 1996; Rogers and Randolph 2000)[5]. In most cases, such predictions cannot readily be tested without a long wait. Given the difficulties of correlative evidence (see below), the heavy reliance on such models of much discussion of the consequences of climate change, in particular, is troubling (Section 5.4).

Typically, more than one, and often several, climatic variables make statistically significant contributions to the explanation of the pattern of occurrence of a species. Much attention has been paid to improving the models used in such studies (this is particularly problematic for fine spatial resolutions, for which environmental data are largely lacking). Of course, hundreds of possible combinations of climatic parameters exist (for temperature alone there are minimum, maximum, mean, and cumulative values for an infinite set of different periods within and between different sets of years), and arguably it would be surprising if one could not be found which described the occurrence of a species reasonably well regardless of any causal relationship. Nonetheless, two observations would seem particularly convincing evidence for the importance of climate in limiting the occurrence of species, both in marine and terrestrial systems.

The first is the existence of outlying populations in what appear to be climatic refugia. For example, Ryrholm (1988, 1989) documents the occurrence of the moth *Idaea dilutaria*, the beetle *Danacea pallipes*, and the spider *Theridion conigerum* on the Kullaberg peninsula in southern Sweden, around 1000 km north of the northern limits of their continuous distributions. The area of the peninsula occupied by these species is distinctly warmer than surrounding, and apparently similar, habitats. This climatic outpost is a result of steep vegetated slopes, the relatively large number of sun hours in spring, proximity to the sea, and the character of the rock outcrop. Artificially created climatic outposts, such as the warm waters generated by nuclear power stations, may operate in a similar way for some species.

Second, there are many examples of species that towards higher latitudes decrease the depths or elevations at which they occur (in terms of climatic gradients, 200 m of elevation is approximately equal to 1° of latitude; Flebbe 1994), increase their height in the intertidal, or which increasingly favour south/north facing slopes (e.g. Fig. 2.4). They thereby effectively maintain the climatic regime that individuals experience, or at least markedly reduce the variation.

Changes in climatic conditions

Broad synchrony between changes in the distributional limits of species and changes in environmental conditions, and particularly climatic ones, has been observed across a spectrum of time scales. For ecological time spans, over which only more

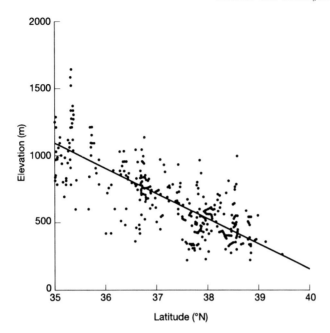

Fig. 2.4 Relationship between the latitude (°N) and elevation of streams in which brook trout *Salvelinus fontinalis* occur in North Carolina and Virginia. From Flebbe (1994).

mobile species have the capacity to respond, three examples will suffice[6]. First, Murawski (1993) regressed the mean and maximum latitude of occurrence of 36 fish and squid species in the northwest Atlantic against average surface- and bottom-water temperatures. Between-year variations in water temperature were significant in explaining changes in mean latitude of occurrence for 12 species in spring and autumn. Responses in maximum latitude to interannual differences in temperatures occurred for five species in both spring and autumn. Similarly, a relationship was found between the maximum altitude at which larvae of a population of spittlebugs, *Neophilaenus lineatus*, occurred in each year of a 10-year period and the weather in March to July in that same year (Fig. 2.5; Whittaker and Tribe 1996). Third, in a detailed analysis, Mehlman (1997) investigated the effects of the unusually harsh winters of the late 1970s on the abundances of three North American passerine bird species. For the Carolina wren *Thryothorus ludovicianus* and the eastern bluebird *Sialia sialis*, changes in proportional abundances were greatest at sites closer to the range edge or with a more severe winter, and sites closest to the edge were at greatest risk of extinction.

El Niño–Southern Oscillation (ENSO) events, in which environmental conditions typically change over large areas, result in the appearance of many marine species at higher latitudes than those at which they normally occur; and some resident species may increase or decrease in recruitment, mortality (perhaps to the point of local, or

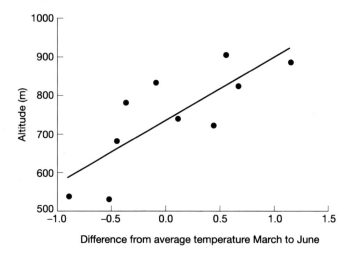

Fig. 2.5 The relationship between the difference from the 25 year mean average temperature from March to June in 1986 to 1995 and the maximum altitude of larvae of the spittlebug *Neophilaenus lineatus* (Auchenorrhyncha, Cercopidae) on Ben Lomond on or about 1 July. From Whittaker and Tribe (1996).

even global, extinction) or abundance (e.g. Brown and Suharson 1990; Hawkins *et al.* 2000; similar observations have been made for terrestrial systems; e.g. Kinnaird and O'Brien 1998; Harrison 2000; Jaksic 2001).

Much attention is at present directed towards determining the extents to which the range limits of species are shifting in response to human-induced global climate change. After controlling for overall population expansions and contractions, Thomas and Lennon (1999) have shown that over a 20-year period the northern margins of many British bird species have moved further north by an average of 18.9 km. Likewise, the ranges of butterflies, and some other groups, have been found to be moving northwards, both in Europe and North America (Parmesan 1996; Cannon 1998; Hill *et al.* 1999; Parmesan *et al.* 1999; Burton 2001).

Over longer periods of time, the distributions of species have ebbed and flowed with shifts in climatic regimes both in marine and terrestrial systems. Burton (1995) argues that since the beginning of the twentieth century changes in climate have probably, at least in part, resulted conservatively in 309 of the 435 European breeding bird species undergoing alterations in their geographic distributions. Of these, 224 species have moved northward, north-westward, and/or westward, and 32 species have retreated south or south-eastward. Between 1900 and 1950, 50 species retreated northwards and since then 55 species have advanced southwards again. Likewise, Burton (2001) states that of 245 species of macrolepidoptera and pyralid moths in Europe whose breeding distributions are known to have altered since 1850, 201 (82 per cent) have expanded their ranges in one or more directions. The vast majority have expanded to the north, north-west, or west, with these changes concerning periods of climatic amelioration. In some cases it is clear that such expansion has entailed dispersal across substantial areas of unsuitable habitat.

The Quaternary record (that for approximately the last 2 million years) shows that migration has been the usual response of organisms to climate change (Huntley 1991, 1994; Bazzaz 1996; FAUNMAP Working Group 1996; Bennett 1997; Shugart 1998; Hewitt 2000). Whilst broad interspecific patterns of movement can be identified, species respond individualistically to climate change—moving in different directions, at different times, and at different rates (Taper *et al.* 1995; Hodkinson 1999; Lawton 2000). For many (most?) areas of the world, on climatic grounds alone, species assemblages would have looked rather different just a few thousand years ago (not that long ago in evolutionary time).

Confidence in the relationship between the distributions of species and climate has extended to the point where the historical distribution of species has been used to reconstruct past patterns of climate (e.g. Coope 1978; Atkinson *et al.* 1987; Coope *et al.* 1998; but for discussion see Andersen 1993, 1996; Coope and Lemdahl 1996); the distribution of plant species has also been advocated as a basis for correcting present-day climate maps (Boyko 1947).

Other evidence

Although suggestive, neither the coincidence of range limits with climatic factors nor the synchronous change in climatic factors and the position of range limits need necessarily imply a direct effect of those factors on the distribution of a species. Aside from the general difficulty of imputing cause and effect from correlation, there are several problems:

1. *Covariation in climatic variables*—variation in many climatic variables is strongly correlated, and thus little can be implied from the coincidence between the range limit of a species and one variable, as this is likely to differ only marginally from the level of coincidence with another facet of climate.

2. *Summary variables*—the summary climatic variables commonly employed in studies of species range limits reflect conditions as measured by a climate station. They will seldom correspond to the conditions that are actually experienced by an organism, which may exploit particular microclimates. Moreover, such summary variables have commonly been corrected to mean sea level and so may not even capture the actual gross climates of areas that are of relevance to flora and fauna.

3. *Indirect effects*—climatic factors which are associated with the range limits of a species may be covarying with or influencing the distribution of resources, competitors, predators, or parasites, to which the distribution of the species is in turn responding. Thus, for example, there is a correlation in Britain between geological structure and climate. The north and west are formed of paleozoic rocks which form extensive uplands that lie across the main pathway of the frontal weather systems from the Atlantic, and which shelter the lowlands of

southeast England that are largely formed of mesozoic and younger rocks and have a more continental climate (Pigott 1970). The distributions of many species follow this discontinuity. Likewise, Cousens and Mortimer (1995) observed that the limits of many weeds in Australia coincide with those of the wheat-growing area, but that it is unclear whether the weeds and wheat share climatic preferences or whether the weeds require this particular habitat management system.

Rather than the coincidence of range limits with climatic factors, better evidence for a direct effect of climate is provided when it can be demonstrated that the climatic conditions *per se* beyond the boundaries of the geographic range of a species are unsuitable. Albeit rather simplistic, Hutchins (1947) distinguished two general ways in which this might occur. Climatic conditions may either impose a boundary by directly killing individuals or they may impose one because they are unsuitable for reproduction or the successful completion of life cycles (arguably, under some circumstances, tantamount also to killing individuals). Both observations were made with particular reference to temperature, which has generally and consistently been considered to be the most important suite of climatic variables limiting the distributions of species (e.g. Merriam 1894; Hutchins 1947; Crowson 1981; Jeffree and Jeffree 1994).

Killing individuals

The imposition of a boundary because climatic conditions kill individuals is doubtless far less frequent than many seem to infer from the general importance of climate in determining range limits. Indeed, the physiological capacities (temperature tolerance, desiccation tolerance, etc.) of many species seem to exceed the range of environmental conditions they are likely, normally, to encounter (Spicer and Gaston 1999). Nonetheless, climate-induced mortality as a determinant of range limits has been demonstrated, or strong evidence has been obtained, in a number of cases (see also Crisp 1965; Willis 1985):

1. *Holly*—low temperatures, particularly when associated with rapid temperature changes, cause a variety of types of injury that are directly or indirectly associated with the freezing of water in plant tissues. These include crown kill, sunscald, winter burn, blackheart and frost cracking, blossom kill, death of vegetative shoots, death of buds and bark, and outright death (Burke *et al.* 1976). Holly *Ilex aquifolium* suffered severe damage and death in Denmark close to its northern limit in mainland Europe during severe winters between 1939 and 1942, apparently caused by frost injury to the cambium (Fig. 2.6; Iversen 1944); the species suffered similarly in Britain during 1963 (Pigott 1970).

2. *Marine bivalve*—over-winter survival of the marine bivalve *Abra tenuis* in the Wadden Sea, at the northern edge of its distribution, is a positive function of winter temperature (Dekker and Beukema 1993). Over two decades, survival was greater in mild than cold winters, and close to zero during all winters that were

Fig. 2.6 Proportions of stems of holly *Ilex aquifolium* killed and injured in Denmark close to its northern limit in mainland Europe during severe winters between 1939 and 1942. Solid shading, killed; grey shading, severely injured; open, slightly injured; filled circle, a killed individual; circle within circle, a severely injured individual; and open circle, a slightly injured individual. From Iversen (1944).

colder than average. Recruitment was also higher in warm than in cold summers, and growth was more rapid in warm than in cold spring–summer periods.

3. *Southern pine beetle*—the northern distribution limit of the southern pine beetle *Dendroctonus frontalis* in the United States approximately matches the isoline corresponding to an annual probability of 0.90 of winter air temperatures reaching less than or equal to $-16°C$ (Ungerer *et al.* 1999). Laboratory measurements of lower lethal temperatures and published records of mortality in wild populations indicate that such temperatures should result in almost 100 per cent mortality.

4. *Painted turtle*—the northern range limit of painted turtles *Chrysemys picta* appears to result from the low temperatures reached in the nests in which hatchlings overwinter in some years exceeding the thermal tolerances of these animals (St Clair and Gregory 1990).

5. *North American birds*—it has been proposed that a physiological ceiling on metabolic rate, preventing the demands of thermoregulation from being met as temperature declines, constrains the northern distributional limits of wintering birds in North America (Root 1988b, 1988c, 1989). However, subsequent analyses have challenged the contention (Repasky 1991), perhaps reinforcing the notion that limits in this case are likely to be determined by the interactions of temperature and biotic factors rather than simply by a metabolic constraint.

Consideration of the role of climate in killing individuals beyond the boundary to a range highlights the potential importance of four things. First, the length of a study may be significant in ascertaining the role of different factors in the occurrence of a species. Extremes of temperature, for example, may be infrequent, and may not be detected by studies of short duration. Second, the rate of climate change may play as major a role in limiting the range of a species as the climatic conditions actually attained. Upon exposure for a moderate period to conditions different from those that they normally encounter, individuals of many species are able to acclimate (with regard to tolerance of low temperatures this is sometimes referred to as 'hardening' in plants) and in so doing attain greater tolerances to extreme events than they would otherwise have been able to express (see Spicer and Gaston 1999 for a review). Likewise, if rates of change are sufficiently slow species may be able to respond behaviourally to avoid their effects. Third, at any location the individuals of a species are likely to differ phenotypically and genotypically, and to exhibit variation in their climatic tolerance, so conditions that kill some individuals may not kill others. Fourth, interaction between climatic extremes and other factors may be significant— for example preventing individuals from feeding, because of the lack of accessible food or because of the need to avoid the extreme climatic conditions.

The above said, the potential for climate and weather extremes to kill organisms is exemplified by the numerous (though, regrettably, scattered in the literature) examples of such events causing mass mortalities of local populations (e.g. Blegvad 1929; Goodbody 1961; Crisp 1965; Horwood and Millner 1998; Cerrano et al. 2000)[7]. For example Kinne (1970) lists numbers of locations at which there have been mass mortalities of marine animals apparently as a result of cold. Many agriculturally important insect pests in the Europe and the United States move northward each summer in their millions and then die, not having formed resident populations capable of surviving winter at high latitudes (Fitt 1989; Cannon 1998). Whether, more generally, such mass mortalities are experienced more frequently towards the range limits of species remains unclear, but seems highly likely.

Limiting reproduction and life cycles

Although less dramatic than killing individuals, more frequently climatic conditions may impose a boundary to the distribution of a species because beyond this point they are unsuitable for reproduction or the successful completion of life cycles (e.g. Barnes 1958; Marshall 1968; MacArthur 1972; Haftorn 1978; Grace 1987;

Cambridge *et al.* 1990; for discussion see also Yom-Tov 1979; Neilson and Wullstein 1983; Elkins 1995). There are many examples, but the following capture much of the diversity:

1. *Crops*—for many agricultural crops, the probability of crop failure, when levels of reproduction of many crops are too low in economic terms, can to a first approximation be predicted for different areas (yielding a risk surface) using accumulated temperature (or the number of degree-days—commonly, the product of the number of days for which mean temperature exceeds an arbitrary standard (usually 0°) and the mean temperature over this period (Parry and Carter 1985; Carter *et al.* 1991a, 1991b)).

2. *Nymphalid butterflies*—four nymphalid butterfly species that share the same primary host plant, the common stinging nettle *Urtica dioica*, have different margins to their geographic ranges in the United Kingdom and Europe (Bryant *et al.* 1997). The distributional margins appear to follow summer isotherms and for three species, small tortoiseshell *Aglais urticae*, peacock *Inachis io* and comma *Polygonia c-album*, differences in relative distribution can broadly be explained in terms of degree-day requirements. The migrant red admiral *Vanessa atalanta* does not fit the predicted pattern, perhaps because it may be more limited by its ability to overwinter. The most northerly species, *A. urticae*, has the lowest degree-day requirement.

3. *Arctic aphid*—the distributional limit of the high Arctic aphid *Acyrthosiphon svalbardicum* in an area of Spitsbergen results from summer thermal conditions, and it survives only at sites with little or no snow cover that have maximal summer thermal budgets, but are exposed to the lowest winter temperatures; laboratory cultures were successfully reared on plant material collected from colder regions than those that the aphid presently occupies (Strathdee and Bale 1995). Arguably, in the high Arctic the distribution of many invertebrates is likely to be defined by the severe thermal conditions alone (Strathdee and Bale 1995).

4. *Geometrid moth*—considering the distribution of the silky wave moth *Idaea dilutaria* in terms of the amount of energy available for development explains some apparently disparate phenomena (Ryrholm 1989). These include the differential impact of different climatic variables in different regions, differences in the type of habitat occupied, and the role of limestone and other edaphic factors.

5. *Freshwater fish*—the northern distributional limits of some freshwater fish can be explained simply by the necessity of the growing season being sufficient that young can complete a minimum amount of growth during their first year of life (Shuter and Post 1990). This is because weight-specific basal metabolism increases as size decreases, with no corresponding increase in energy storage capacity. Thus smaller fish tend to be less tolerant of starvation conditions, which typify the long period of cold temperatures characteristic of winter in temperate zone lakes.

6. *Swans*—the whooper swan *Cygnus cygnus* takes 130 days to complete its breeding cycle. Within that part of its range north of the Arctic circle only half the summers are ice-free for this length of time, and at these latitudes breeding is often unsuccessful (Elkins 1995, p.83). Its smaller relative, the Bewick's swan *Cygnus columbianus* can breed further north, since it has a cycle of only 100–110 days.

In addressing the restrictions on reproduction and completion of life cycles imposed by climatic conditions, there is again a need to ensure consideration of processes operating on an appropriate time frame. Thus, some plant species fail to set seed, or suffer other forms of failed or reduced sexual reproduction, at their range limits and survive by vegetative reproduction, or by longevity (until conditions change; Salisbury 1926, 1932; Pigott 1970, 1975, 1992; Pigott and Huntley 1981)[8]; it should be noted that in some species individual reproductive performance is a positive function of population size (e.g. Ågren 1996; Fischer and Matthies 1998) and peripheral populations may be smaller (Section 4.5). The match between observed limits and limits of reproduction is likely to be closer for species in which individuals have shorter longevity (Woodward 1997). This is emphasized by studies of the small-leafed lime *Tilia cordata*; more is probably known about the determinants of the geographic limits of this species than for any other. This long-lived tree is estimated to have reached the British Isles about 7000 BP (Huntley and Birks 1983), and to have reached its northern limit in this region between 7000 and 5000 BP (Pigott and Huntley 1981). The present-day distribution extends to this same northern limit; however, the present-day reproductive limit lies some 200 km further south, raising the possibility that its continued presence at the northern limit is a result of climate conditions from long ago (Pigott 1989; Woodward 1990). The present-day reproductive limit of *T. cordata* is caused by the temperatures required for pollen tubes to grow at a sufficient rate to enable fertilization to occur (Pigott and Huntley 1981). At its northern limit in Finland, with its relatively continental climate, temperatures are high enough to allow fertilization, but subsequently fall and may not be sufficient to allow complete development of the embryo and endosperm (Pigott 1981).

The role of climate

The argument for the importance of climate in the direct limitation of the ranges of species has sometimes been put very forcefully. Thus, Brett (1956, p.80) observed that 'There can be no doubt that the lethal temperature exerts limiting effects on the geographic distribution and freedom for successful existence...'. However, even the demonstration that individuals cannot survive or complete life cycles under climatic conditions beyond the edge of the range of a species remains in one sense only indirect evidence that those conditions are limiting. In evolutionary time, if other factors limit the occurrence of a species, then the potential to survive, develop, and successfully reproduce under conditions beyond these limits may be lost (especially if their maintenance is costly in resources), driving a tendency toward coincidence between conditions that are experienced and conditions which can be tolerated. Because physiologies must at least be sufficient to tolerate conditions within the range, this may result in systematic interspecific patterns of geographic variation in physiologies and the positions of geographic limits being observed in ecological time. Thus, Brattstrom's (1968) observation that the LT_{50} cold lethals of species of anurans (those at which 50 per cent of individuals die) decrease with the latitude of their northern-most occurrence (Fig. 2.7) may be suggestive of a role for temperature in range limitation, but equally it may imply no such thing (although in the particular case of northern hemisphere anurans at high latitudes there is little evidence that other factors are actually limiting). Of course, regardless of how any match between physiology and distribution is driven (and this could rapidly degenerate into a 'chicken or egg?' argument), it may provide a basis for predicting the occurrence of a species. Recognition that climatic conditions will tend to limit species distributions through physiological processes has led some to model occurrences based on variables that are most likely to represent distinct physiological limiting factors (e.g. Sykes *et al.* 1996).

Climate undoubtedly plays an important role in limiting the geographic ranges of many species. What is less clear is the extent to which that role is a primary or a secondary one with respect to other factors.

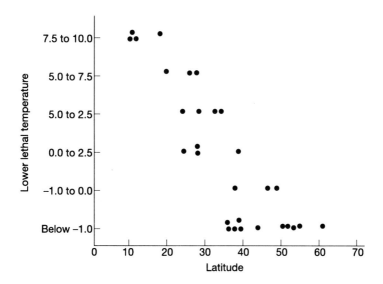

Fig. 2.7 Relationship between the lower lethal temperature ($LT_{50}/24$ h) (°C) and the latitude (°N) of the northern-most occurrence of species of anuran amphibians. Each dot represents 2–25 individuals. From Brattstrom (1968).

2.2.3 Other abiotic factors and habitat

Whilst it may be able to account for the distributions of many, climate is not sufficient to account for the distribution of potentially large numbers of species. As MacArthur (1972) observes, if this were not true then botanical and zoological gardens, with representatives from a diversity of climatic conditions, would be impossible. In such gardens, other environmental conditions are often modified to suit particular species (e.g. provision of resources), competitors and predators are eliminated, and diseases are strictly controlled. What proportion of species is strictly limited by climate remains unknown. It is not really even very clear whether some groups are more responsive to climatic conditions *per se* than are others; amongst ecologists, impressions as to such differences abound, but seem little more developed than that.

Many abiotic factors beside climate may restrict the geographic ranges of species. The variety of factors that could potentially play such a role is probably limited as much by the imagination of the investigator as by anything else. Possibilities include, depending on the system, aspect, slope, light, CO_2, topography, geology, soils, water, salinity, pH, trace metals, fire, and other forms of disturbance (although some of these are clearly related to climate). Thus, the red pine *Pinus resinosa* is restricted to islands and shores of lakes at its northern limit in Quebec, probably because of the abundance of sites protected from lethal fires (Bergeron and Gagnon 1987; Bergeron and Brisson 1990).

Climate, other abiotic factors, and perhaps biotic factors too, tend to be integrated in the habitat requirements (used in the broadest sense) of species. At the edge of their range, many species tend to be restricted to quite specific habitats and microhabitats (Carter and Prince 1985a; Kavanagh and Kellman 1986; Thomas 1993; Strathdee and Bale 1995; Yagami and Goto 2000; but see Diekmann and Lawesson 1999 for a counter-example)[9]. Such a pattern might potentially be explained simply as a 'sampling effect', if the abundances of species also decline towards their range limits (Section 4.5); the smaller number of individuals would be expected by chance to occur in a smaller number of habitats, and predominantly in the more widely distributed habitats which the species can exploit. However, those habitats occupied at the range edge may be atypical of those occupied elsewhere in the geographic range (Griggs 1914; Carter and Prince 1985a; Kauhala 1995; Marren 1999; Yagami and Goto 2000; Jones *et al.* 2001). For example, red spruce *Picea rubens* has been found at its lower elevational range limit growing as small disjunct populations in bogs, a habitat unusual for the species elsewhere in its range (Webb *et al.* 1993). Moreover, many of the habitat shifts that occur towards range edges are systematic, such as occurrence on south facing slopes in the northern hemisphere and north facing ones in the southern hemisphere. The occupation of specific microhabitats is particularly obvious in biogeographic transition zones, which may often exhibit high levels of species richness as a result of the diversity of microhabitats they embrace and the numbers of species at their range limits.

The occupation of unusual habitats at range limits may reflect the absence of more typical habitat, changes in conditions that render more typical habitat difficult or impossible to occupy, or changes in conditions that render otherwise unusual habitats more favourable.

2.2.4 Interspecific interactions — consumers

In some cases, the limits to the occurrence of species may, foremost, be imposed by the occurrence of the particular resources that they consume. The simplest case, although it is still not necessarily a simple one, is perhaps that of the specialist consumer or parasite, whose distribution must be contained within that of its host. The majority of such species, and probably all, do not occupy the entire distribution of their host, and their range limits only coincide with those of the host in some regions, if anywhere (e.g. Pepper 1938; Gilbert 1980; Strathdee and Bale 1995; Quinn *et al.* 1997a, 1998; Ungerer *et al.* 1999; Ménendez and Thomas 2000; but see Gutiérrez and Thomas 2000)[10]. For example at a resolution of 50×50 km, host-specific species of macrolepidoptera occupy between 0.014 and 0.901 of the geographic distributions of their host plants in Britain, with an average of *c.* 0.4 (Quinn *et al.* 1997a). The eight monophagous moths attacking common reed *Phragmites australis*, occupy between 0.014 and 0.671 of the distribution of this plant (Fig. 2.8).

The failure of specialist consumers to occupy the entire geographic range of hosts could occur for reasons to do with the host, because of other limits on occurrence, or perhaps some combination of the two. The host may become too scarce towards its range limits for viable populations of the consumer to persist or individuals of the host towards it range limits may provide unsuitable resources for the consumer. However, commonly the limits of specialist consumers lie so far within those of the host that such effects seem unlikely, and it is the response of the consumer to other environmental factors that seems to prevent their exploiting the full geographic range of the host.

If specialist consumers are discriminating, and the mere presence of a host is not indicative of its suitability, the situation becomes yet more complex with non-specialists, which may exhibit changes in resource use from one area to another. Here again, consumers seldom occupy the entire range of a single host and, commonly, different hosts are used in different areas of the geographic range of a herbivore or parasite, and different prey in the case of predators (e.g. Fox and Morrow 1981; Strong *et al.* 1984; Clay *et al.* 1985; Pekkarinen and Teräs 1993; Hughes 2000). Few studies seem to have explored just why this should be so, beyond issues of resource availability. Many herbivorous insects become more specialist in their host use towards their range limits. For example the swallowtail butterfly *Papilio machaon* is restricted to the scarce, but locally abundant, milk-parsley *Peucedanum palustre* in Britain, but will use a variety of plants in mainland Europe. Hodkinson (1997) found that there was a progressive reduction of host plant species and tissues exploited by the willow psyllid *Cacopsylla groenlandica* towards high latitudes in Greenland, such that towards the northern limit of its range it became

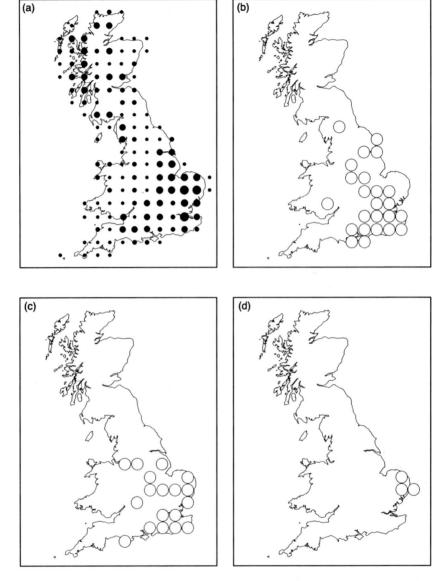

Fig. 2.8 Distribution, at scale of 50×50 km, of (a) common reed *Phragmites australis* (symbol size reflects the number of 10×10 km squares from which the plant has been recorded), and the eight monophagous moth species feeding on it in Britain, (b) obscure wainscot *Mythimna obsoleta*, (c) flame wainscot *Senta flammea*, (d) Fenn's wainscot *Photedes brevilinea*, (e) twin-spotted wainscot *Archanara geminipuncta*, (f) brown-veined wainscot *Archanara dissoluta*, (g) white-mantled wainscot *Archanara neurica*, (h) large wainscot *Rhizedra lutosa*, and (i) fen wainscot *Arenstola phragmitidis*. From Quinn *et al.* (1997a).

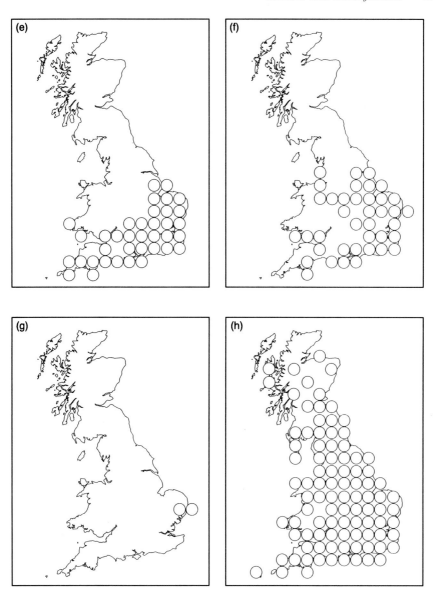

Fig. 2.8 *(continued)*

highly specialized, feeding only on female catkins of one host despite the presence of alternative hosts.

The diversity of hosts in an area may determine the likelihood that a consumer can occur there. Thus, Koenig and Haydock (1999) find that the effective

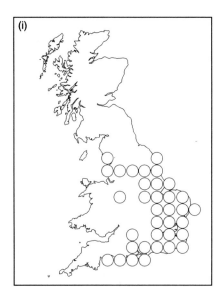

Fig. 2.8 *(continued)*

distributional limit of the acorn woodpecker *Melanerpes formicivorus* in the south-western United States and along the Pacific coast is set by sites where the diversity of oaks (genera *Quercus* and *Lithocarpus*) drops to a single species and not by the limit to the distribution of oaks. This seems to result from declines in the variability in overall acorn production and the probability of acorn crop failure with increasing number of oak species (due to asynchrony in acorn production between different species of oaks).

The disappearance of Hume's leaf warbler *Phylloscopus humei* at its northern range limit in the Indian subcontinent appears to be associated with a decline in arthropods in trees towards the north (Gross and Price 2000). This is, in turn, associated with a decline in the numbers of leaves on trees, which may be a consequence of the influence of climate on leaf retention. Such limits on resource availability, perhaps mediated by climate, may commonly play a role in restricting the geographic ranges of species.

2.2.5 Interspecific interactions—competitors

As well as providing resources, other species may act as competitors with, or consumers of, a particular species. In either of these capacities, they may ultimately limit its geographic range, by preventing expansion of its boundaries (Darwin 1859). In the main, the interactions are not sufficiently strong for geographic limitation to result (although some have argued that the enormous benefits experienced by some introduced species coincident with their escape from native competitors and natural enemies indicate that these strongly influence distributions). However, Davis *et al.* (1998a, 1998b; see also Jenkinson *et al.* 1996) have shown how interactions can be potentially very important influences on the distributions of species across simple environmental gradients, using an elegant set of microcosm experiments, a model assemblage of three fruitfly species, *Drosophila melanogaster*, *D. simulans* and *D. subobscura*, and a parasitoid wasp *Leptopilina boulardi*. Temperature clines were generated by linking eight cages in series and housing each pair of cages in separate incubators at different temperatures. Competitive interactions between the fruitfly species altered their distributions and abundances from those found in single-species experiments. Adding the parasitoid further modified these distributions and abundances.

Interspecific competition has been considered to influence the spatial distributions of species across a broad range of spatial scales (Connor and Bowers 1987). The idea that it plays an important role in determining the limits to the geographic ranges of species is an old and persistent one (Griggs 1914). In principle, it is easy to see that competition for resources may prevent the coexistence of two or more species, limiting their geographic spread where the two meet. However, studies have remained essentially correlational and testing, typically for pairs of species, for abutting range limits and non-random patterns of overlap in distributions, or inferring competition from the occurrence of parapatry (where ranges are separate, but slightly overlapping).

Bullock *et al.* (2000) demonstrate that there are strong negative associations between the occurrence of two closely related species of dwarf gorse, a perennial shrub, *Ulex minor* and *U. gallii*, at three geographic scales, across the British Isles and France, across the heaths of Dorset, southern England, and on individual heaths. Indeed, apparent co-occurrences detected at coarser spatial resolutions largely disappeared when the species were mapped at finer resolutions. The authors conclude that the distributions of these two species are not independent, that they cannot coexist, and therefore that their ranges are limited by competition.

The scale of analysis may generally prove important in the conclusions that are reached as to the evidence for limitation of occurrences by competition. Pielou (1977a; see also Pielou 1978) tested whether, because of competitive exclusion among them, the latitudinal spans of congeneric seaweed species overlap less than expected. She found no evidence for competition at this scale, whilst noting that at the scale of individual localities zonation is common, apparently as a result of competition. This emphasizes the more general point that the determinants of range limits that are most apparent may vary with spatial resolution (Carter and Prince 1988).

Most studies are conducted at a single spatial resolution. Letcher *et al.* (1994) examined the amount of overlap in the geographic ranges of species of Palaearctic and British mammals. Closely related species tended to overlap more than expected by chance, suggesting that the distributions of species are constrained more by their niche requirements than by competition. Species that overlapped less than expected by chance were more similar in body size, and vice versa, suggesting that

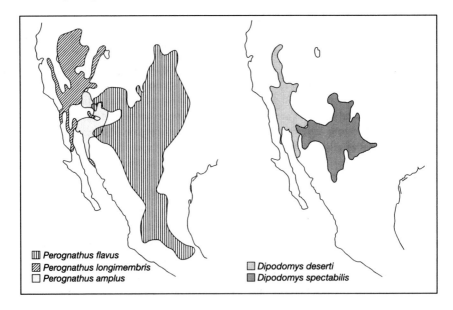

Fig. 2.9 Geographic ranges of small pocket mice (*Perognathus* spp., <11 g) and of large kangaroo rats (*Dipodomys* spp., >100 g). Species of similar body size exhibit small overlap in their ranges. From Bowers and Brown (1982).

competition may cause species with similar body size to reduce range overlap (see also Blackburn *et al.* 1998a). Bowers and Brown (1982) found a similar result for granivorous desert rodents in North America, with species of similar body size (mass ratios <1.5) overlapping less in their geographic distributions than expected by chance; an explanation based on allopatric speciation was rejected because many of the non-random associations were not between closely related species (Fig. 2.9).

Hersteinsson and Macdonald (1992) provide evidence that the northern limit of the range of the red fox *Vulpes vulpes* is determined by resource availability (and ultimately by climate), but the southern limit of the arctic fox *Alopex lagopus* (which is smaller-bodied and less resource demanding) is determined by the distribution and abundance of the red fox. They argue that interspecific competition between other pairs of canid species may similarly determine range limits.

Ferrer *et al.* (1991) suggest that changes in the breeding ranges of starlings in the Iberian peninsula over the past 30 years provide a good example of competition acting as a limiting factor in the dispersal of species. Both the European starling *Sturnus vulgaris* and the spotless starling *Sturnus unicolor* have expanded their ranges in the region over this period, at low rates. Where the species have come into contact, sympatric areas have appeared which have slowed down the rate of expansion of both species. The area of sympatry has stopped expanding, perhaps indicating that colonization of new areas is blocked by the presence of the other species.

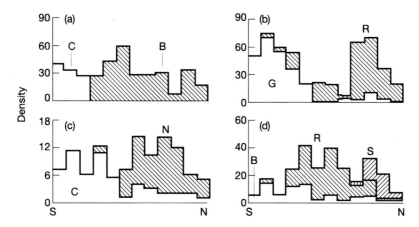

Fig. 2.10 Patterns of density (birds per km²) variation along a latitudinal gradient (1200 km, through relatively uniform bottomland deciduous forest in middle North America) for selected pairs of phylogenetically or ecologically related land-bird species. (a) Carolina chickadee *Poecile sclateri* (C) and black-capped chickadee *Poecile atricapillus* (B), (b) gnatcatcher *Polioptila caerulea* (G) and redstart *Setophaga ruticilla* (R), (c) cuckoos (C) and northern oriole *Icterus galbula* (N), and (d) red-bellied woodpecker *Melanerpes carolinus* (B), red-headed woodpecker *Melanerpes erythrocephalus* (R), and sapsuckers (S). From Emlen *et al.* (1986).

In some cases, evidence of a role for competition in limiting ranges has been drawn from the observation of opposing spatial gradients in the densities of pairs of species. Thus, Emlen *et al.* (1986) report gradients of this form amongst some pairs of probably competing bird species, such that where one species was abundant the other was more scarce and vice versa (Fig. 2.10).

Other arguments for competitive limitation have been yet more inferential. For example Loehle (1998) finds a trade-off between height growth rate and freezing tolerance for 22 species of North American trees and argues that, as a result, at their southern margins northern hemisphere trees are outcompeted by species with a faster growth rate. A role for competition is often also inferred when the absence of a species from a region cannot be attributed to its physiology. Thus, competition with a congeneric species was suggested as restricting the distribution of the southern African ice rat *Otomys sloggetti*, when its absence below altitudes of 2000 m was found not to be determined by its thermal physiology (Richter *et al.* 1997). *O. irroratus* and *O. sloggetti* have allopatric distributions, and Richter *et al.* (1997) argue that competition excludes the latter from exploiting lower altitudes, whilst thermoregulatory considerations preclude the former from exploiting higher ones. This combined effect of temperature and competition in determining species distributions is, they propose, a widespread one. Importantly, the coincidence of range limits with particular sets of climatic conditions cannot be used to reject a role for limitation by competition. Because climatic conditions are likely to be important in determining the relative competitive abilities of species, range limits may exhibit such coincidence.

Studies that also demonstrate experimentally that competition is limiting occurrences on broad scales are seldom performed (although such experiments are frequently conducted to examine small-scale competitive interactions). However, Jaeger (1970, 1971) showed, using information on microgeographic distributions and exclusion experiments, that the narrowly distributed salamander *Plethodon richmondi shenandoah* (found only in three isolated talus slopes in the Blue Ridge Mountains of western Virginia) is largely restricted to areas of talus on Hawksbill Mountain, Shenandoah National Park, Virginia because it is excluded from soil outside the talus by *P. cinereus* (Fig. 2.11). The talus is, however, suboptimal habitat for *P. r. shenandoah*, and if *P. cinereus* is artificially excluded the survivorship of *P. r. shenandoah* is significantly better in soil.

MacArthur (1972) argued that the southern limits of many northern hemisphere species are determined by competitive interactions. Terborgh (1985) attributes the elevational limits of two-thirds of the species in his study of Andean birds to competitive exclusion (direct or diffuse) and one-sixth to ecotones. He further argues that evidence from temperate regions suggests, in comparison, a much greater importance of ecotones and a much reduced importance of competition, probably as a result of more drastic structural changes across ecotonal boundaries.

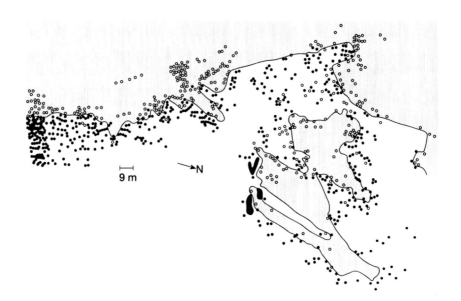

Fig. 2.11 The distributions of collected *Plethodon richmondi shenandoah* (open circles) and *P. cinereus cinereus* (filled circles) salamanders, on a portion of the talus-soil interface on Hawksbill Mountain, Virginia. Stippled areas are talus, the white area is deep soil, and solid lines indicate the interface. Large dark objects are large boulders. From Jaeger (1970).

2.2.6 Interspecific interactions—the consumed

Although the possibility that they play a role has long been entertained, predation and parasitism have often been held, in themselves, to be unlikely determinants of the range limits of prey or host species. This, it has been argued, is because although a predator or parasite may depress the abundance of its prey or host to very low abundances in an area, it is unlikely to drive it extinct before either it becomes extinct itself (because it cannot sustain a sufficient population) or switches to an alternative, more abundant resource.

However, this position is coming under increasing pressure. Concentrating on host–parasitoid dynamics, Hochberg and Ives (1999) developed a simple mathematical model that shows that a natural enemy can enforce a limit. Some important features of this model are that host net reproduction is assumed to vary systematically across its range as a modified Gaussian distribution (see Section 4.5.1), and that the two most significant conditions for natural enemy-enforced range limits are that (i) the theoretical host equilibrium density in the presence of the natural enemy be very small at sites eliminated from the host's range, and (ii) the natural enemy disperses at high rates. However, the authors also observe that there are two other ways in which a natural enemy could cause a clear boundary to the range of its prey. One would be if there was a sharp cut-off to the prey's reproductive rate, and the proportion of hosts that was invulnerable to attack was positively correlated with this rate. The other would be where a predator could limit the density or drive to local extinction one of two (or more) alternative prey species in regions where they co-occurred.

The last of these seems the most likely scenario for range limitation by natural enemies. Indeed, such 'apparent competition' may be a significant force in community structure more broadly (Holt 1977; Holt and Lawton 1994). If natural enemies are shared between two species, then formally, in general terms, the consequences for these two species are identical to forms of interspecific competition for limiting resources (Williamson 1957). Thus, the introduced citrus psyllid *Trioza erytraea* was driven extinct on the island of Réunion by the introduced parasitoid *Tetrastichus dryi*, which could also exploit the alternative host *Trioza eastopi* (Samways 1994; for other examples see Nafus 1993).

Price *et al.* (1988) argue that there is evidence that species with a wide geographic range tend to displace species with a narrower range when a parasite common to both has differential impact, because wide-ranging species have a greater pool of parasites (Section 3.5.6; see also Pagel *et al.* 1991; Tompkins *et al.* 2000). Certainly, there is growing evidence that shared parasites have the potential to explain cases where the abundance and distribution of one host have declined whilst those of another have increased (Rushton *et al.* 2000; Tompkins *et al.* 2000).

Other 'predator–prey' interactions have also been argued, on occasion, to act as a factor influencing the distributions of prey species, including herbivory (gopher herbivory on aspen *Populus tremuloides*, Cantor and Whitham 1989; sheep herbivory on the crucifer *Moricandia moricandioides*, Gómez 1996; slug herbivory on the composite *Arnica montana*, Bruelheide and Scheidel 1999; ungulate herbivory on Arizona willow *Salix arizonica*, Maschiniski 2001) and conventional predation. Thus, Brönmark and Edenhamn (1994) provide evidence that the reproductive success and occurrence of the tree frog *Hyla arborea* in the most northerly region of its distribution is influenced by the presence of fish, probably through predation on tadpoles and possibly also on eggs. Pienkowski (1984) concludes that predation probably determines the southern nesting limits of the ringed plover *Charadrius hiaticula* in Europe, and possibly other bird species. The occurrence of large carnivores may limit the westerly distribution of megapodes (Megapodidae, Aves), which by their mound-building behaviour seem particularly vulnerable to predation (Dekker 1989).

Exclusion by predators or parasites may be more likely in species with limited dispersal abilities, which cannot readily take advantage of the subsequent absence of those natural enemies.

This said, it appears that it is not uncommon for species to escape the influence of at least some of their natural enemies in peripheral populations (Radomski *et al.* 1991; Hodkinson 1999 and references therein). Thus, for example, in temperate regions at range edges the populations of herbivorous insects seem frequently to be found to be free, or relatively free, of parasitoids. Thus, the holly leaf-miner *Phytomyza ilicis* suffers high levels of mortality throughout much of its natural geographic range as a result of larval parasitism, but towards its northern limit although pupal parasitism occurs larval parasitism drops to zero (Brewer and Gaston, in press). In the case of *Phytomyza ilicis*, the lack of larval parasitism at high latitudes may result from the absence of an alternative host for the larval parasitoid *Chrysocaris gemma*.

Of course, the best, and perhaps most convincing, examples of the constraints on the ranges of species that can be imposed by a competitor or predator concern the impacts of humans on other species. The present, highly elevated global species extinction rates reflect the human capacity to carry limitation of the occurrence of a species to its conclusion, namely to extirpate it from its entire geographic range. Much of this range limitation is indirect, through environmental change, resource destruction, and the accidental or intentional introduction of other competitors and predators. This nicely exemplifies the complexities of attributing causation to the determinants of the geographic boundaries to the ranges of species.

The relative importance of abiotic and biotic factors in determining range limits has been much debated (e.g. Darwin 1859; MacArthur 1972; Miller *et al.* 1991; Davis *et al.* 1998a, 1998b; Hodkinson 1999; Gross and Price 2000)[11]. Two divergent viewpoints can be distinguished. On the one hand, at the edge of its geographic range, the stresses caused by abiotic factors can be seen as potentially reducing the competitive ability of a species, and increasing its vulnerability to predation or its susceptibility to pathogens. Thus, biotic factors may become important in limitation. On the other hand, at the edge of its range abiotic factors may commonly render a species too scarce for biotic factors to have any marked effect.

2.2.7 Multiple factors

Some authors have argued that the ranges of individual species typically are limited by very few factors (abiotic or biotic), and perhaps only by one (e.g. Twomey 1936). This is a gross oversimplification (Salisbury 1926). First, in any particular area, and at a given time, the distributional limit of a species may be mediated by complex interactions between factors. Unfortunately, the effects of multiple interacting factors on the limits to species distributions are difficult to study, at least when more is required than simply entering multiple variables into a regression analysis.

Second, given the temporal variation in the state of very many abiotic and biotic factors, in any one area the identities of the factors and complexes of factors limiting the range of a species may change through time.

Third, the identities of limiting factors and complexes of factors may change from one part of a geographic range to another (e.g. Barnes 1958; Taylor 1977; Bullock *et al.* 2000; Gross and Price 2000). Indeed, spatial variation in the factors limiting the distribution of a species is almost certainly the norm, and constancy the exception (Hutchins 1947). Thus, some part of the geographic ranges of many species is limited by physical boundaries (e.g. coasts), whilst limitation elsewhere is caused by other factors.

Of course, not only does the environmental template vary in space, but so also may the characteristics of a species itself. The geographic range sizes of most species are much larger than are the typical dispersal distances of individuals or gametes (see Chapter 3), and thus despite such movements they typically differ phenotypically (with regard to their morphologies, physiologies, life histories, ecologies, etc.) and genotypically across their geographic ranges (e.g. Moore 1949; Endler 1977; Loik and Nobel 1993; Ayres and Scriber 1994; van Herrewege and David 1997; Elmes *et al.* 1999)[12]. Indeed, such variation may be necessary if they are to attain large geographic range sizes (and it has become a major issue in the context of sourcing individuals for reintroduction programmes for conservation; Section 5.6). In short, the individuals of a species are very likely, in some sense, simply not to be the same 'beast' from one part of the species range to another.

A valuable way of beginning to disentangle the effects of different factors in limiting the geographic ranges of species is to carry out transplant experiments, introducing populations to sites beyond the bounds to its extent of occurrence (Salisbury 1932; taking care about the provenance of the introduced individuals, to address the complications of geographic phenotypic and genotypic variation). Here, responses should be enhanced, and therefore the relevant factors easier to identify; although, of course, if limitation is very variable in space and if transplant sites are rather extreme the results may prove to be misleading. Such experiments remain relatively scarce, and certainly more so than reciprocal transplants between sites within the geographic ranges of individual species. Nonetheless, a number of studies have been performed, most commonly for plants (which have the obvious advantage of not requiring to be caged to limit any subsequent movement), although several of these experiments have concerned internal rather than external range boundaries (Table 2.1). Most seem to conclude that whilst initial transplants may

Table 2.1 The outcomes of experiments that have transplanted individuals of species beyond the limits to their geographic ranges

Species	Outcome	Source
Plants		
Gambel oak *Quercus gambelii*	Most survived 1st year; only 6% survived beyond two summers	Neilson and Wullstein (1983)
Brown algae *Fucus serratus*	Successfully grew and reproduced (similar growth and reproduction rates to those transplanted within normal range)	Arrontes (1993)
Cheatgrass *Bromus tectorum*	Successfully grew and reproduced, but low level of seed production	Pierson and Mack (1990)
Barley *Hordeum vulgare*	Grew successfully at two northern sites outside main growing area in UK but ears took longer to develop and yield lower at upland site	Prince (1976)
Grass sp. *Hordeum murinum*	Grew and reproduced successfully in 1st year, poor establishment in 2nd year and reproduction declined until none germinated by 4th year	Davison (1977)
Prickly lettuce *Lactuca serriola*	Successfully grew, reproduced and persisted	Prince and Carter (1985)
Annual phlox *Phlox drummondii*	Successfully grew and reproduced, albeit at a reduced level	Levin and Clay (1984)
Vervain *Verbena officinalis*	Grew and reproduced at some sites (died out at others); poor germination at high altitude/low temperature sites	Woodward (1990)
Wall pennywort *Umbilicus rupestris*	Successfully grew and reproduced	Woodward (1990)
Round-leaved fluellen *Kickxia spuria*	Successfully grew, reproduced and persisted	Carter and Prince (1981)
Grass vetchling *Lathyrus nissolia*	Successfully grew, reproduced and persisted	Carter and Prince (1981)
Stonecrop sp. *Sedum rosea*	Successfully grew and reproduced at elevations below usual range	Woodward and Pigott (1975)
Orpine *Sedum telephium*	Successfully grew and reproduced at elevations above usual range	Woodward and Pigott (1975)
Animals		
Thimbleberry aphid *Masonaphis maxima*	Successfully reproduced but new generation failed to produce sexual progeny and population failed to persist	Gilbert (1980)
Baltic clam *Macoma balthica*	Some survived at least 10 months after translocation (not known if reproduced), but subsequent absence may be due to emigration or mortality; weight-index lower and respiration rate greater than non-translocated individuals	Hummel *et al.* (2000)

grow and successfully reproduce, levels of reproduction tend to be impaired, and the long-term persistence of populations is much less likely.

The practical difficulties associated with determining the relative roles of factors in limiting the geographic range of a species in any one place, and the likelihood that different factors are important in different places, greatly complicate the prediction of the likely response of species' distributions to future climate and other environmental change. It is clear that species will move in different directions and at different rates, but more detailed statements are commonly problematic and always make the questionable assumption that future distributions will reflect present responses to environmental conditions. This is significant, because such predictions are likely to be increasingly important for management and conservation practices (Section 5.4).

2.3 Population dynamics

Whether abiotic or biotic, factors that limit the distributions of species, whether operating singly or in combination, can be viewed as generating this limitation by influencing the vital rates of populations.

2.3.1 Single populations

Put most simply, the number of individuals in a local population is determined by the number of births (b), deaths (d), immigrants (i) and emigrants (e), such that

$$N_{t+1} = N_t + b - d + i - e$$

where N_t is the population at time t and N_{t+1} is the population at time t + 1. In the long term, a species will occur in an area when

$$b - d + i - e \geq 0$$

although if the population size is large enough, or individuals persist for long periods, this inequality may not be met at some, perhaps prolonged, times. The range edge therefore represents the point at which this inequality no longer holds for sufficient periods. This circumstance may obviously be attained in a wide variety of ways—births and/or immigration may become too small, deaths and/or emigration may become too large, or some combination of the two. None, some, or all of b, d, i, and e may be functions of population size (i.e. are density dependent), although in all likelihood at least some will be (Watkinson 1985). The interaction between density-dependent and density-independent processes determines the actual size of a population (Williamson 1972). Only if some or all of b, d, i, and e are density dependent will population size remain more or less constant from year to year and it can be said that population size is regulated.

Analysis of models and empirical data have confirmed that small variations in levels of births, deaths, immigration, and emigration may be sufficient to produce abrupt limits to ranges. This means that although the position of a range limit may be obvious, detecting the variation in population parameters that brings it about may in practice be difficult (Carter and Prince 1985a), even more so given likely errors in their empirical measurement. This is particularly true where population size is determined by an interaction between a rate that is non-linearly related to population size and density-independent processes (Fig. 2.12; Watkinson 1985), as may commonly be the case. This means that even small differences in abiotic or biotic factors, and that may otherwise appear insignificant, may be crucial in determining range limits if they influence a vital rate. This may, at least in part, explain why field studies documenting the demographic reasons for range limits are scarce, and predominantly concern plant species (for which there is more likelihood of measuring vital rates with some accuracy than for perhaps most animal species).

In an important, though much neglected, paper Beddington *et al.* (1976) propose that towards range edges several things could occur: (i) changing environmental conditions alter average population parameters such that no stable population dynamics exist; (ii) unpredictable events attain levels such that the probability of a population being pushed to extinction tends to unity within a few generations; or (iii) the time taken by a population to return to equilibrium levels becomes so long that small perturbations to population size lead to extinction.

A number of authors have argued that towards the edges of geographic ranges, local populations will exhibit dynamics that are differentially (compared with other local populations) dominated by abiotic factors or experience less density-dependent regulation (Chiang 1961; Mayr 1963, 1970; Huffaker and Messenger 1964; Richards and Southwood 1968; Whittaker 1971; Coulson and Whittaker 1978; Bengtsson 1993;

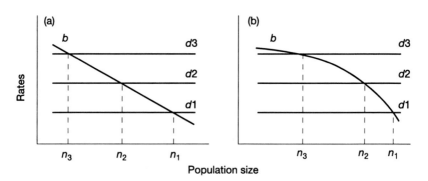

Fig. 2.12 The change in equilibrium population size (*n*) that results from a change in the density-independent death rate (*d*) when the density-dependent birth rate (*b*) is (a) linear and (b) non-linear. From Watkinson (1985).

Thomas *et al.* 1994). There is only very limited empirical evidence with which to assess whether this is actually so. However:

1. *Tick*—weaker density dependence of mortality towards the range edge has been found for the tick *Rhipicephalus appendiculatus* in Africa (Randolph 1997). Here, less favourable abiotic conditions cause greater deviations from the density-dependent relationship.

2. *Cercopid*—in Britain, nymphal, adult, and generation mortality of the cercopid bug *Neophilaenus lineatus* has been found not to be detectably density dependent at a high elevation site near the edge of its range, but nymphal mortality is inversely density dependent, and adult and generation mortality directly density dependent at a lowland site well within the range of the species (Whittaker 1971; see also Coulson and Whittaker 1978).

3. *Clonal plant species I*—comparison of the dynamics of two clonal plant species, the divaricate sunflower *Helianthus divaricatus* and fragrant sumac *Rhus aromatica* (a shrub), at their northern range limit and further south (closer to their range centres) revealed that for both species, the northern populations showed a larger temporal and spatial variation in some of their vital rates and in annual population growth rates (Nantel and Gagnon 1999).

4. *Clonal plant species II*—a reciprocal transplant experiment with the clonal aquatic plant greater spearwort *Ranunculus lingua* found that ramets from marginal and central populations performed equally well in the marginal habitat, but ramets from the marginal population suffered a greater reduction in most growth parameters in the central habitat (Johansson 1994). Ramets from the marginal population always produced more but smaller rhizomes than those from the central population. In the central habitat, small ramets were more likely to be diseased and damaged by grazing. In the marginal habitat mortality was largely density independent and arose from abiotic factors, favouring many small vegetative offspring.

Although such dynamics might occur for a variety of reasons (e.g. Gaston and McArdle 1994; Wilson *et al.* 1996), if there is less density-dependent regulation towards range limits then one might perhaps expect to see greater temporal variability in the size of more peripheral populations. A number of studies have claimed to find support for this proposition. Thus, the relative variability of local populations of species of North American sparrows has been found to increase towards their range limits, as their local abundances decline (Curnutt *et al.* 1996). Likewise, Thomas *et al.* (1994) found increased fluctuations in the sizes of populations towards the northern edges of the ranges of butterfly species in Britain. The populations of caribou *Rangifer tarandus* and the muskox *Ovibos moschatus* in the high Arctic, close to the limits of their ability to adapt to the extreme conditions, are characterized by wide fluctuations in numbers often culminating in local extirpation (Klein 1999). With specific reference to flatfishes, Miller *et al.* (1991) hypothesized that recruitment variation should, as a consequence of difference in the significance of abiotic and biotic factors, be higher at the northern edge of the range of a (northern hemisphere) species, least near the centre of the range, and intermediate near the southern limit. Some evidence has been consistent with this pattern and some has not (Myers 1991; Rijnsdorp *et al.* 1992; Leggett and Frank 1997), with the suggestion being made that in some cases predictions regarding changes in population dynamics towards range limits may be nullified by broader latitudinal trends in population variability (Henderson and Seaby 1999).

A logical extension of the argument that peripheral populations will exhibit greater temporal variability in size is that they will also tend to experience a higher frequency of local extinctions. Indeed, species at the edge of their geographic ranges have often been hypothesized to be at greater risk of extinction (Terborgh and Winter 1980). This is what is commonly observed (e.g. Whittaker 1971; Dekker and Beukema 1993; Kattan *et al.* 1994; Christiansen and Pitter 1997; Marren 1999; but see Gilbert 1980). Nathan *et al.* (1996) record that between 1863 and 1993, 204 bird species were recorded as breeders in Israel, and that of the 185 species that bred regularly, 14 species have become extinct and 58 species are threatened. The extinct species were significantly more likely to be at the periphery of their geographic ranges in Israel, and the non-threatened species were significantly less likely to be at their periphery in this region. In countries where a significant proportion of species attain their range limits (e.g. Canada), these species contribute heavily to levels of extirpation and local population losses.

2.3.2 Multiple populations

Consideration of levels of local extinction is based on the recognition that species typically comprise multiple populations even at their range limits. Indeed, multiple populations and metapopulation structure may be more likely at range limits, where abundance structures may be more fragmented (Section 4.5). Single populations may have limited persistence, in which case the range edge must be determined as the point at which the rate of establishment of populations minus the rate of extinction falls below zero. Carter and Prince (1981) explored this using a model identical in form to the general deterministic epidemic model used in the mathematical theory of infectious diseases. If a is the number of unoccupied colonizable sites available (susceptible) to a species, o is the number of colonizable sites that are occupied (infective), β is the rate at which colonizable sites become occupied (the infection rate), and δ is the rate at which colonized sites cease to be occupied (the removal rate), then the rate at which colonizable sites become colonized is

$$do / dt = \beta ao - \delta o$$

For the species to survive in an area

$$do / dt \geq 0$$

which requires that

$$a \geq \delta / \beta$$

Again, such a model enforces the idea that small changes in limiting factors can determine the position of a range edge (Carter and Prince 1981, 1988). Towards the range limit, the number of susceptible sites, infection rate, and removal rate may all change, eventually shifting the parameters of the threshold condition such that it is no longer satisfied. Individual organisms might be able to survive, but the metapopulation would not.

Holt and Keitt (2000) use the Levins metapopulation model (Levins 1968, 1969) to illustrate how gradients in occupiable sites, colonization rate, and extinction rate can each generate limits (see also Lennon *et al.* 1997). The basic model can be written as

$$\frac{dn}{dt} = cn(k-n) - en$$

where of all the patches available in a landscape a fraction k are potentially suitable for occupation by a species, a fraction n are actually occupied, each occupied patch goes extinct at a constant per patch rate e, and each empty patch is colonized at a constant per patch rate c. If we assume that there is a series of landscapes in which the dynamics of a species can be described by the above equation, and that these landscapes lie along a smooth environmental gradient characterized by a single spatial dimension x, then the model can be rewritten to capture this broader scale, by making its parameters functions of the position of a landscape with regard to x

$$\frac{dn(x)}{dt} = n(x)[k(x) - n(x)]c(x) - e(x)n(x)$$

In this model no dispersal occurs between different landscapes, but this will make little difference to the general results provided typical dispersal distances are small relative to the size of a landscape.

A species will be absent from all points along the gradient x for which

$$k(x) < \frac{e(x)}{c(x)}$$

If the fraction of potentially occupiable sites (k) declines along the gradient, then even holding extinction rate (e) and colonization rate (c) constant for all values of x, a range limit will eventually result. This means it is possible that a limit may be attained even if the conditions within patches remain constant approaching that limit (Carter and Prince 1981). Moreover, for a given colonization rate, a species with a lower extinction rate will be able to survive where there are relatively fewer potentially occupiable sites, and will therefore be able to persist closer to the point at which, say, climate prevents an individual organism from growing (Carter and Prince 1988).

A range limit may equally be met if the extinction rate rises across the environmental gradient, but the proportion of potentially occupiable patches and the colonization rate remains constant. Many of the factors discussed earlier in this chapter could plausibly lead to an increase in extinction rates, particularly if they result in a reduction in local population sizes along a gradient (see Section 4.4). As Holt and Keitt (2000) observe, this increase is, however, likely to mean that the conditions within patches are rather different moving towards the range limit.

Equally, a range limit will result if the colonization rate declines across the environmental gradient, but the proportion of potentially occupiable patches and the extinction rate remain constant. Holt and Keitt (2000) recognize two major ways in which the colonization rate might decline. First, if local abundances decline, then the number of potential colonists is likely to do so. Second, environmental conditions along the gradient may progressively reduce successful dispersal between patches. This is tantamount to the range of a species being dispersal limited.

In practice, of course, variation in any or all of the proportion of potentially occupiable patches, colonization rate, and extinction rate could give rise to range limits in systems for which the scenario modelled by Holt and Keitt (2000) is reasonably realistic. Observing systematic variation in any one of these parameters towards range limits will not be sufficient to indicate what is causing range limitation.

2.4 Genetics

In considering limitations on the geographic occurrence of a species, as well as asking how abiotic or biotic factors may cause limitation, and how population dynamics may change towards limits, the question arises as to what prevents traits from evolving so that an organism can expand its range beyond its observed bounds (Mayr 1963). This is a particular case of the more general problem of what limits natural selection. Plainly the potential for such evolution exists, in as much as, for example, selective breeding has been employed to develop crop varieties that are able to tolerate shorter growing periods and lower temperatures and thus be grown at higher latitudes than was previously possible. Several hypotheses have been proposed to explain what prevents such evolution from occurring in most species (Hoffmann and Blows 1994; Hoffman and Parsons 1997; see also Haldane 1956; Antonovics 1976). The first three concern low heritabilities of traits determining range boundaries in marginal populations:

1. *Low overall levels of genetic variation occur in marginal populations because of small population size.* This may prevent range expansion, simply by reducing the

likelihood that the required genes occur in marginal populations. Marginal populations are expected to have low genetic variation if they are subject to intense stabilizing selection, if they are small and therefore more prone to inbreeding and genetic drift, and if they frequently go extinct and recolonization takes place by only a few individuals. However, the maintenance of genetic variation in marginal populations, even if these are scattered and small, may result if species have particular life history characteristics, such as being long-lived, outcrossing, and having substantial gene flow. Levels of variation could also be higher in marginal populations if selection pressures are more heterogeneous. There are obvious links here with the earlier discussion of local population dynamics at range limits (Section 2.3.1).

2. *Traits show low heritability as a consequence of directional selection in marginal environments.* This could occur because, for example, a species is close to its physiological limit at its range margin.

3. *Traits show low heritability because of environmental variability in marginal environments.* Heritability is determined by genetic and environmental variance, thus an increase in the variability of environmental conditions can act to reduce heritability. Such variability may also hinder the expression of differences between genotypes. The occurrence of greater variability in abundances towards range edges (Section 2.3.1) would provide one piece of suggestive evidence that environmental variability is indeed greater in marginal areas.

A further three hypotheses concern genetic interactions between traits:

4. *Changes in several independent characters are required for range expansion, and so favoured genotypes occur too rarely.* Many environmental parameters exhibit spatial covariance, which means that expansion of the distribution of a species beyond an existing boundary may necessitate simultaneous genetic changes in a number of traits. This may be difficult to achieve.

5. *Genetic trade-offs between fitness in 'favourable' and 'stressful' environments prevent the increase of genotypes adapted to stressful conditions.* If the conditions that limit the geographic ranges of species occur only occasionally in time, then the genes that would enable those ranges to expand may only be favoured for insufficient periods for them to increase in frequency, and may be costly outside those periods.

6. *Genetic trade-offs among fitness traits in marginal conditions prevent traits from evolving.*

Two other hypotheses are that:

7. *The accumulation of mutations that are deleterious under stressful conditions prevents adaptation.* This may occur if deleterious mutations are only expressed under particularly stressful conditions, but when expressed their deleterious effects prevent range expansion (Hoffman and Parsons 1997).

8. *Genotypes favoured in marginal populations are swamped by gene flow from central populations.* If gene flow is sufficient, then the occurrence at range edges of alleles that would enable range expansion may be swamped by alleles from populations elsewhere. Gene flow is likely to be predominantly into peripheral populations if populations more central to ranges are larger, pushing the former away from local optimum adaptation (Section 4.5). This possibility has been modelled by Kirkpatrick and Barton (1997), who found that under some conditions a small parameter change can dramatically shift the balance between gene flow and local adaptation. Case and Taper (2000) have extended the Kirkpatrick and Barton model to include interspecific competition and the frequency-dependent selection that it gives rise to. This shows that in the absence of environmental gradients or barriers to dispersal, competition will not limit species ranges at evolutionary equilibrium. But, such limits can arise from the interaction between interspecific competition, environmental gradients, and gene flow.

Overall, the empirical evidence in support, or otherwise, of these mechanisms is rather limited, and it is unclear which are the most significant and under what circumstances. It is extraordinary that such a fundamental question has received so little attention. Most of the available data concern the extent of genetic variation toward the limits of geographic ranges, and concern non-quantitative variation (e.g. allozymes, molecular variation), begging the question of what would be found with quantitative variation as the two may be poorly correlated. Whilst some studies do show evidence of lower levels of genetic variation in marginal populations (Fig. 2.13) others do not (Table 2.2), and, indeed, Safriel *et al.* (1994) report the obverse pattern, with a cline of increasing genetic variability from central to peripheral populations of the chukar

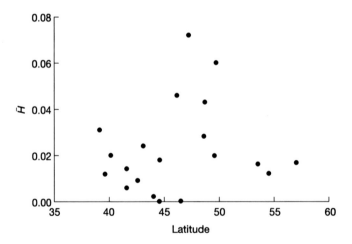

Fig. 2.13 Variation in average heterozygosity per population ($\bar{H}$) calculated from allele frequency data, with latitude (°N) for the western North American spotted frog complex *Rana pretiosa*. From Green *et al.* (1996).

Table 2.2 Studies comparing the genetic variability of core and peripheral populations

Species	Variation < in peripheral populations		Source
Gymnosperms			
Lawson cypress *Chamaecyparis lawsoniana*	√		Millar and Marshall (1991)
Lodgepole pine *Pinus contorta*	√		Yeh and Layton (1979), Cwynar and MacDonald (1987)
Jeffrey pine *Pinus jeffreyi*	√		Furnier and Adams (1986)
Ponderosa pine *Pinus ponderosa*	X	no consistent pattern (greater variability at some loci but not others)	Hamrick *et al.* (1989)
Pinyon pine *Pinus edulis*	X	no difference between central and marginal populations	Betancourt *et al.* (1991)
Pitch pine *Pinus rigida*	√	but effect weak	Guries and Ledig (1982)
Norway spruce *Picea abies*	X	less variability in central European populations compared to N and E Europe	Lagercrantz and Ryman (1990)
Norway spruce *Picea abies*	X	no difference between populations	Tigerstedt (1973)
Douglas fir *Pseudotsuga menziesii*	√		Li and Adams (1989)
Douglas fir *Pseudotsuga menziesii*	X	no difference between populations	Yeh and O'Malley (1980)
Angiosperms			
River birch *Betula nigra*	√		Coyle *et al.* (1982)
Honey locust *Gleditisia triacanthes*	√		Schnabel and Hamrick (1990)

(continued)

Table 2.2 *(continued)*

Species	Variation < in peripheral populations		Source
Wild tomato *Lycopersicon pimpinellifolium*	√		Rick *et al.* (1977)
[primrose] *Lysimachia volkensii*	√		Agnew (1968)
Pitcher plant *Sarracenia pupurea*	√		Schwaegerle and Schaal (1979)
[tea sp.] *Camellia japonica*	X	no systematic difference between populations	Wendel and Parks (1985)
[tea sp.] *Camellia japonica*	X	greater variability at N periphery of range (Korea) than in populations to S (Japan)	Chung and Chung (2000)
[Phlox] *Phlox spp.*	X		Levin (1977, 1978)
[Lythraceae] *Decodon verticillatus*	√		Eckert and Barrett (1993)
Thale cress *Arabidopsis thaliana*	√		Kuittenen *et al.* (1997)
Strapwort *Corrigiola littoralis*	√		Durka (1999)
Sticky catchfly *Lychnis viscaria*	√		Lammi *et al.* (1999)
Manuka *Leptospermum scoparium*	X	no consistent pattern in variability	Wilson *et al.* (1991)
Swamp pink *Helonias bullata*	X	variability greatest in S and lowest in N of range	Godt *et al.* (1995)
Invertebrates			
[bivalve] *Elliptio complanata*	√		Kat (1982)
[bivalve] *Lampsilis radiata*	√		Kat (1982)
[land snail] *Isognomostoma isognomostoma*	X	no systematic difference between populations	Van Riel *et al.* (2001)

(continued)

Table 2.2 *(continued)*

Species	Variation < in peripheral populations		Source
Baltic clam *Macoma balthica*	?	variability decreases toward S limit, but increases toward N limit	Hummel *et al.* (1997)
[Sea urchin] *Echinometra*	√		Palumbi *et al.* (1997)
[drosophilid flies] *Drosophila spp.* (5 spp.)	X	none of studies suggest a reduction in heterozygosity in marginal populations	Brussard (1984)
Sand scorpion *Paruroctonus mesaensis*	√		Yamashita and Polis (1995)
Clouded apollo butterfly *Parnassius mnemosyne*	?	variability lowest in S and higher towards central and N of range (disjunct populations)	Descimon and Napolitano (1993)
Vertebrates			
Spotted frog *Rana pretiosa*	√		Green *et al.* (1996)
European tree-frog *Hyla arborea*	√		Edenhamn *et al.* (2000)
Chukar *Alectoris chukar*	X	higher variability at S periphery	Safriel *et al.* (1994)
Great reed warbler *Acrocephalus arundinaceus*	√		Bensch and Hasselquist (1999)
Greenfinch *Carduelis chloris*	X	greater variability in S populations	Merilä *et al.* (1997)
Stephens kangaroo rat *Dipodomys stephensi*	√		Metcalf *et al.* (2001)

partridge *Alectoris chukar*. Likewise, whilst population size and genetic variation may often be positively correlated (Ellstrand and Elam 1993; Frankham 1996; Weidema *et al.* 1996; Fischer and Matthies 1998; Jones *et al.* 2001) they are far from always so (Järvinen *et al.* 1976; Ellstrand and Elam 1993; van Rossum *et al.* 1997; Menges and Dolan 1998; Kahmen and Poschlod 2000; Wolf *et al.* 2000)[13]. Unfortunately, insufficient attention has been paid to date to the spatial scales at which different studies have been conducted. It seems likely that, at least in the north temperate regions where many have been performed, much of the broadest scale spatial pattern of genetic variation reflects the history of range expansion and contraction (often associated with glacial periods; e.g. Sage and Wolff 1986; Green *et al.* 1996; Soltis *et al.* 1997; McCusker *et al.* 2000; Schmitt and Seitz 2001; Zeisset and Beebee 2001). Here, genetic variation commonly (although not exclusively) declines with distance away from glacial refuges, which often lie in what are now the southern regions (and perhaps margins) of the distributions of the species, suggestive of founder effects associated with colonization/recolonization. How readily this translates into patterns of variation at, close to, and a moderate distance from the range limits remains unclear. Similarly, different patterns of genetic variation towards range limits may be explained by differences in the more recent histories of species, particularly between common species in decline, often as a result of human activities, and species with a naturally restricted distribution (Luijten *et al.* 2000).

Most attention seems, at present, to be directed to the role of gene flow preventing local adaptation and subsequent range expansion (hypothesis 8). However, it seems likely that in many species gene flow into peripheral populations is simply too small for such a mechanism to be operating.

Whatever the processes operating at the genetic level that prevent species from further expanding their geographic range limits, they seem to be quite stable ones. Whilst there are ample examples in both the geological and historical records of species which have undergone range expansions in response to changes in the availability of appropriate environmental conditions, there seem to be very few where species have undergone range expansions without such changes, implying that evolutionary modifications that enable species to expand into new sets of environmental conditions are rare events (Eldredge 1999). Lewontin and Birch (1966) document a possible case of range expansion through the generation of a novel physiology. The fruitfly *Dacus tryoni* has, since at least 1860, extended its geographic range southward in eastern Australia, with a concomitant increase in altitudinal range. Lewontin and Birch suggest that this broadening of geographical range was previously limited by climatic factors rather than biotic ones. Laboratory investigations revealed that range expansion could be correlated with wider heat and cold tolerances, with the genetic variation required for selection apparently being provided by introgression from a second species of *Dacus*, *D. neohumeralis*. Unfortunately, hybridization of the two *Dacus* species in the laboratory resulted in the production of populations better adapted to only some of the experimental temperatures used.

2.5 In conclusion

The determination of how and why the geographic ranges of species are limited should be a central objective of ecological research, given that ecology is commonly defined as the study of the abundance and distribution of organisms. This objective continues to constitute a major challenge, in large part because of the multiple levels at which it can be addressed. The vast majority of studies to date have concerned explanations at just one of these levels, be it the abiotic and/or biotic factors that prevent further spread (inevitably actually some rather small subset of the possible factors), the population dynamics that prevent persistence, or the genetic mechanisms that cause geographic restriction. Thus, the literature provides good empirical examples of these different kinds of explanations, but rather little insight into how they relate to one another, and almost no comprehensive understanding of why any given species occurs where it does and not elsewhere.

From the mass of theoretical and empirical studies, it is possible nonetheless to extract some more general messages that may withstand the test of time:

1. The factors that limit the geographic range of a given species at a particular place are often difficult to predict, without detailed knowledge of the biology of the species and the prevailing (and perhaps historical) environmental conditions. Not infrequently, the important factors will differ even between two closely related species in the same area. Indeed, no two species, however closely related and likely to exhibit similar biologies, have identical geographic distributions—most sister species show quite marked differences in their distributions.

2. Physical factors are limiting to at least some parts of the geographic ranges of a high proportion of species.

3. At a global scale most species are dispersal limited, there being areas in which they could persist but which they cannot reach, often in other biogeographic regions.

4. Both abiotic and biotic factors are important in limiting species' geographic ranges. Any obsession with the role of climatic factors in isolation seems likely to prove misguided and, in general, observational studies cannot satisfactorily discriminate abiotic and biotic influences.

5. Climatic factors seem likely to be more important in limiting ranges through constraints on development times and opportunities for successful reproduction than through their effects on the mortality of adult individuals.

6. There is no intrinsic reason why interspecific interactions cannot limit the geographic ranges of species. Indeed, such limitation is probably quite common but may involve complex sets of interspecific effects, perhaps involving multiple species.

7. The interactions between available habitat and colonization and extinction rates that prevent the persistence of metapopulations at range limits may be subtle, and the importance of small changes in one or more of these may therefore be difficult to detect.

8. The importance of past events on the observed limits to the geographic ranges of species increases with the longevity of individuals (which may become established under different conditions) and their dispersal abilities (which may determine how closely distributions track changes in environmental conditions).

3

Range size

Limited range, especially for species and genera, is, on the whole, a much more general phenomenon than wide distribution.

R. Hesse *et al.* (1937)

3.1 Introduction

Whatever the factors responsible, the overall degree of constraint on the geographic ranges of species is extremely variable. In consequence, the areas of these ranges differ enormously; Brown *et al.* (1996) suggest interspecific variation of 12 orders of magnitude, which would be similar to the variation observed amongst species in local densities (Damuth 1987). At one extreme lie those species with either very few individuals, in some cases only one known specimen, or constrained to very small habitat patches (often mountain tops or oceanic islands). Examples include (i) many species of Proteaceae in the Cape Floristic kingdom, which are restricted to one or two populations covering total areas of less than 5 km² (Tansley 1988); (ii) several cave-adapted arthropods that are endemic to four small, old limestone towers that occur within a 2 km² area near Chillagoe, Queensland, Australia, some of which only occur in single towers within the group (Hoch and Howarth 1989); (iii) the Devil's Hole pupfish *Cyprinodon diabolis*, argued by some to be the most restricted extant vertebrate, whose entire global population occupies an area of only 200 m² (Soltz and Naiman 1978); (iv) the Gough Island bunting *Rowettia goughensis*, which is endemic to the remote Gough Island (total area 65 km²) in the South Atlantic and which occupies only limited parts thereof; and (v) the bandro *Hepalemur griseus alaotrensis*, a lemur restricted to reedbeds by Lake Alaotra, Madagascar.

Some of the most restricted 'species', particularly of plants, may be hybrids. Thus, for example, Rossetto *et al.* (1997) find that *Eucalyptus graniticola*, known from a single plant located on a granite outcrop south-east of Perth in Australia, is a hybrid of *E. rudis* and *E. drummondii*. In other cases, the severe restriction of the ranges of species is not artefactual in this sense, small geographic ranges are natural and appear to have persisted for a long period (e.g. many 'island' species). In many others, they are attributable, in part at least, to human activities. Thus, for example, many species in the New Zealand herpetofauna are restricted to a few isolated locations, remnants of once wider distributions, primarily as a result of introduced mammals (Towns and Daugherty 1994). In some groups and regions, such as land birds in Polynesia, range reductions have been the norm (Steadman 1995), leading to instances of pseudoendemism where species have been limited to much smaller areas than their natural distributions (Grayson 2001). The species listed in Red Data Books are heavily dominated by the last of these three groups, those whose distributions have been influenced by human activities, their high risks of extinction resulting in major part from past, and on-going, reductions in their historical geographic ranges (Section 3.4.3; e.g. IUCN 1996; Oldfield *et al.* 1998).

At the other extreme of variation in the sizes of geographic ranges lie the 'cosmopolitan' species, although no species has been found to be genuinely cosmopolitan. Probably the most widespread species are marine, with many being distributed throughout much of the oceans (Hesse *et al.* 1937). These include some species of algae, foraminifera (Gooday *et al.* 1998; Gooday 1999), jellyfish (e.g. *Aurelia aurita*), fish (e.g. some shark and tuna species), seabirds (based on foraging ranges, e.g. sooty shearwater *Puffinus griseus*, Wilson's storm-petrel *Oceanites oceanicus*), and several cetaceans (e.g. sperm whale *Physeter macrocephalus*, killer whale *Orcinus orca*, bottlenose dolphin *Tursiops truncatus*).

In the freshwater and terrestrial realms, notably widespread species include probably many micro-organisms (including some human diseases; Fenchel 1993; Finlay *et al.* 1999), plants such as knotgrass *Polygonum aviculare*, shepherd's purse *Capsella bursa-pastoris*, fat hen *Chenopodium album*, common chickweed *Stellaria media*, annual meadow grass *Poa annua*, and bracken *Pteridium aquilinum* (Conquillat 1951); butterflies such as painted lady *Cynthia cardui*, long-tailed blue *Lampides boeticus*, and monarch *Danaus chrysippus*; birds such as great white egret *Egretta alba*, cattle egret *Bubulcus ibis*, green-backed heron *Butorides striatus*, osprey *Pandion haliaetus*, peregrine *Falco peregrinus*, common moorhen *Gallinula chloropus*, black-winged stilt *Himantopus himantopus*, and barn owl *Tyto alba*; and mammals such as house mouse *Mus domesticus/musculus*, ship rat *Rattus rattus*, common rat *Rattus norvegicus*, red fox *Vulpes vulpes*, gray wolf *Canis lupus* (argued by some to have the greatest natural range of any terrestrial mammal other than humans; Ville de la *et al.* 1998), weasel *Mustela nivalis*, and red deer *Cervus elaphus*, and, of course *Homo sapiens* which is probably the most widely distributed of terrestrial vertebrates (as well as its terrestrial occurrence, the human species now, at any one moment, also has individuals dispersed across the surface of the oceans in ships and on marine platforms, in the oceans in diving suits and submersibles, through the atmosphere in aeroplanes, and often in space).

In less well known groups, the apparently large ranges of some species may simply be an artefact of poor or inadequate taxonomy, hiding the existence of cryptic species (homonymy) (e.g. Knowlton 1993; Etter *et al.* 1999; Klautau *et al.* 1999; Martin and Bermingham 2000; Murray and Dickman 2000; Shaw 2001). Multiple species are considered as one. This may particularly be the case for groups that lack complex morphological characters that could be useful in systematic analyses, such as some marine invertebrates, or where such characters are under strong stabilizing selection or developmental canalization. Equally, however, levels of synonymy of species in some groups increase with geographic range size (Gaston 1996b), suggesting that some genuinely widespread species are at present falsely regarded as comprising two or more species. For example the numbers of synonyms associated with valid names of species of *Enicospilus*, ophionine ichneumonid wasps, are, on average, far greater for those with geographic range extents of more than 5000 km than for others (Table 3.1; Gaston 1996b). Homonymy is, on balance, probably the more powerful force at present, resulting in a tendency for the average geographic ranges of species in given taxonomic groups to be overestimated.

In several cases, species have very large geographic ranges because they have been intentionally or accidentally introduced into areas to which previously they were not indigenous; some of these are food sources (e.g. chicken, cow), some provide services (e.g. honeybees *Apis mellifera* as pollinators), and some are serious weeds, pests, or diseases (the global scale of introductions is vast; Ebenhard 1988; Ruiz *et al.* 1997; Pimentel *et al.* 2000). Amongst the plants, the world's worst weed has been argued to be nutgrass *Cyperus rotundus*, a sedge native to India, and reported to be a weed in 52 crops in 92 countries (Holm *et al.* 1977). Amongst the animals, the house mouse *Mus domesticus* is one of the best known examples. Its original range spread from Nepal westwards to North Africa, and western and southern Europe, but it is now found in the Americas, Australasia, and southeast Africa, as well as on many islands (Corbet and Harris 1991). From samples of agricultural pests (almost certainly biased toward those which are more widespread), Rapoport (2000) reported that 52 out of 260 insect species have been recorded from six biogeographic regions (Palaearctic, Ethiopian, Nearctic, Neotropical, Oriental, Australian), 90 out of 264 phytopathogens, and 18 out of 334 weed species.

Equally, other species would have had extensive ranges had it not been for reductions caused, directly or indirectly, by human influence. Thus, many of the

Table 3.1 The extents of the geographic ranges of *Enicospilus* (Ophioninae, Ichneumonidae, Hymenoptera) wasp species in the Indo-Papuan region, and the numbers of synonyms associated with each valid specific name; data from Gauld and Mitchell (1981)

Range (km)	Number of synonyms											
	0	1	2	3	4	5	6	7	8	9	10	11
1–999	134											
1000–1999	32	2										
2000–2999	18		1									
3000–3999	12	5	1									
4000–4999	11		2									
5000+	16	15	9	8	3	1		1	2		1	1

Table 3.2 Present and historical geographic ranges of some large mammalian carnivores, from Farlow (1993); fossil evidence suggests that in earlier times the distributions of some of these species may have been yet broader (e.g. Turner and Antón 1997)

Species	Geographic range
lion *Panthera leo*	Balkans and Arabia to central India, nearly all of Africa
tiger *P. tigris*	Much of Eurasia
leopard *P. pardus*	Much of Africa and Eurasia
jaguar *P. onca*	Southern United States to northern Argentina
snow leopard *P. uncia*	Mountainous areas from Afghanistan to Lake Baikal and eastern Tibet
cheetah *Acinonyx jubatus*	Middle East to central India, Africa except for the central Sahara and rainforests
cougar *Felis concolor*	Most of North America to southern Chile and Patagonia
spotted hyaena *Crocuta crocuta*	Sub-Saharan Africa except in rain forests
coyote *Canis latrans*	Most of North America
gray wolf *C. lupus*	Most of Eurasia and North America
hunting dog *Lycaon pictus*	Most of Africa
asiatic black bear *Ursus thibetanus*	Most of central and eastern Asia
American black bear *U. americanus*	Most of North America
brown bear *U. arctos*	Most of Eurasia (except tropical regions), northern Africa, most of North America
polar bear *U. maritimus*	Arctic Eurasia and North America

largest-bodied, extant species of carnivorous mammal originally had, but only in some cases still retain, continent-wide or intercontinental geographic ranges, often occurring under environmental conditions markedly different from those to which they are now confined (Table 3.2). Likewise, the range sizes of many species in Australia have undergone marked declines since the arrival of Europeans (Fig. 3.1). Any correlations between range limits and climatic factors would not in these cases provide any evidence of causality.

The rest of this chapter addresses this great variation in the sizes of the geographic ranges of species in detail, particularly with regard to the form this variation takes. The chapter divides into two main sections. The first examines the shape of the

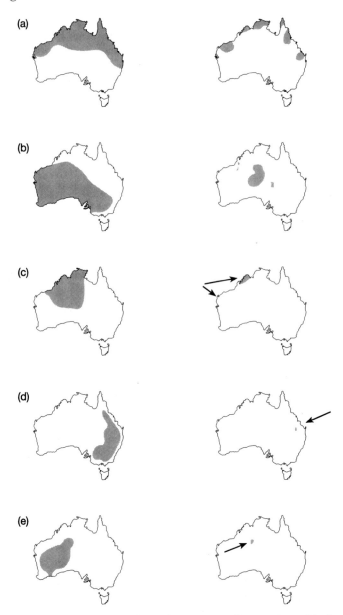

Fig. 3.1 Geographic range contractions of some Australian species, with distribution in 1788 on left and at present on right, for (a) northern quoll *Satanellus hallucatus*, (b) rabbit-eared bandicoot *Macrotis leucura*, (c) golden bandicoot *Isodon auratus*, (d) nail-tailed wallaby *Onychogalea fraenata*, (e) western hare wallaby *Lagorchestes hirsutus*, (f) trout cod *Maccullochella macquariensis*, (g) burrowing bettong *Bettongia lesueur*, (h) numbat *Myrmecobius fasciatus*, (i) pig-footed bandicoot *Chaeropus ecaudatus*, and (j) thylacine *Thylacinus cynocephalus*. From the collation of Burgman and Lindenmayer (1998).

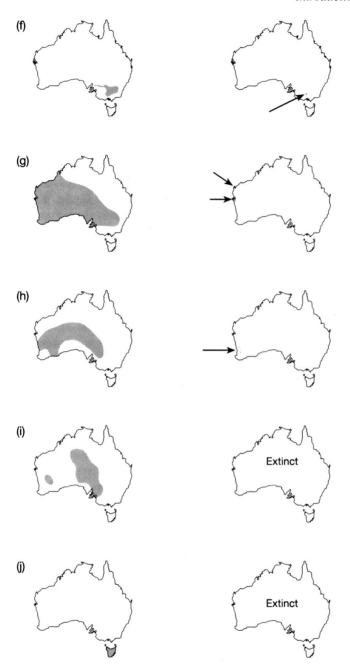

Fig. 3.1 *(continued)*

species–range size distribution, the frequency distribution of the numbers of species with ranges of different sizes, and its determinants. The second section examines patterns in, and correlates of, geographic range sizes. A shorter concluding section addresses the factors determining the size of the range of a species. At the outset, however, it is important to clarify what is meant by the size of a geographic range.

3.2 Extents of occurrence and areas of occupancy

Two broad proximate objectives in the measurement of geographic range sizes can be distinguished. These are to quantify either the area within the outer most limits to the occurrence of a species, or the area over which the species is actually found. These two quantities have been termed, respectively, the *extent of occurrence* and the *area of occupancy* of a species (Gaston 1991a, 1994b). The latter will tend to be consistently smaller, because species do not occupy all areas (or habitats) within the geographic limits to their occurrence (no species is continually distributed in space). Thus, Roberts and Hawkins (1999) observe that among coral reef fishes only 0.34 per cent of the average extent of occurrence consists of coral reefs, and for most species the area of suitable habitat will be much less. At first, this strikes one as a very extreme difference. However, if measured at a fine resolution, area of occupancy may often comprise only a tiny proportion of extents of occurrence. Because both extent of occurrence and area of occupancy can be measured in many ways, such differences need not always be evident.

The extent of occurrence of a species tends to be most relevant in the context of broad issues of biogeography. Uncertainty in the measurement of this quantity is chiefly associated with the need to determine the position of the outermost limits to occurrence, particularly given the structural complexity of range limits and their movement through time (Sections 2.1, 2.2.2). Disagreements often centre on the treatment of extreme outlying populations, which may disproportionately influence the magnitude of extents of occurrence and whose inclusion within this extent will result in the incorporation of substantial continuous areas in which the species plainly does not and will never occur (in the extreme, for example, species with distributions that have major discontinuities or disjunctions, these may be areas of ocean for terrestrial species, and areas of land for oceanic ones). Given that extents of occurrence are generally used to provide a very broad brush view of range sizes such problems are, however, commonly not of sufficient magnitude to be of great concern. Indeed, in the absence of better information, macroecological studies commonly assume that the distribution of a species is continuous between the extreme limits to its occurrence (e.g. Pielou 1977a, 1978; Stevens 1989; Gofas 1998; Roy *et al.* 1998; Clarke and Lidgard 2000), although numbers of species are known to have significant discontinuities and at fine resolutions there are numerous 'holes' in distributions (Section 4.5).

The area of occupancy of a species tends to be more relevant as a measure of geographic range size in the context of many macroecological, and macroevolutionary and applied questions, where what is required is an estimate of the area over which a species is actually distributed, regardless of how widely separated its constituent populations may be. The difficulty with its application is that the actual area that is measured rests fundamentally on the spatial resolution at which occupancy is recorded (Erickson 1945; Gaston 1991a, 1994a; Maurer 1994; Pearman 1997; Kunin 1998; Cowley *et al*. 1999). The finer the resolution the larger the area from which the species is found to be absent, and the smaller the area over which it is found to occur. At the limit, at the finest resolution and measured instantaneously, the area of occupancy of a species would be the sum of the areas occupied by all individuals in the global population. In practice, this is neither a practicable nor a particularly useful measure, what is required is something more reflective of the way in which a species uses space on the scale of multiple seasons or years and at a spatial resolution at which it is tractable to score presence and absence.

It has been suggested, on a number of occasions, that the reduction in the area of occupancy of a species with progressively finer and finer resolutions of mapping may mean that the spatial distributions of species are fractal or approximately so (see Williamson and Lawton 1991; Kendal 1992; Virkkala 1993; Maurer 1994; Kunin 1998; Finlayson 1999). In fact, it seems unlikely that they are strictly fractal, in the sense that the structural details of area remain constant across the full breadth of spatial resolutions; doubts have been expressed as to the evidence and likelihood that the geometry of nature in general is fractal (Avnir *et al*. 1998). Even if they were fractal, whether this could satisfactorily be demonstrated for ranges is doubtful because of the tendency for sampling efficiency to decline with finer resolutions or for marked extrapolations (or interpolations) to be required to compensate for this effect (Gaston 1994a). What is perhaps more important is the recognition that there is not a single spatial resolution at which the area of occupancy of a species can or should be measured. There are innumerable such resolutions, and they will give different answers. In making interspecific comparisons, and the like, it is therefore necessary to ensure adequate comparability in the way in which this is done. In considering individual species, the resolution at which area of occupancy is measured has proven to be particularly significant when evaluating temporal changes in the ranges of species. Coarser resolutions of mapping underestimate rates of decline in areas of occupancy, with the extinction of local populations potentially being undetected at even rather fine resolutions (e.g. 10×10 or 2×2 km; Hodgson 1991; Ehrlich 1994; Thomas and Abery 1995; Asher *et al*. 2001). Thus, Thomas and Abery (1995) found that for 12 species of butterfly of intermediate rarity in the county of Hertfordshire, a 2×2 km grid resulted in estimates of decline that were, on average, 35 per cent higher than estimates based on a 10×10 km grid. In a similar fashion, the spread of species may not be detected at coarse resolutions, where substantial increases in area of occupancy may occur without increasing the number of mapping units that are observed to be occupied. This appears to be taking place in some regions of the United Kingdom where the range boundaries of species are moving in response to climate change, but this movement remains too limited to be detected at the resolution employed by regional atlases.

Various approaches have been suggested for the measurement of area of occupancy (Gaston 1991a, 1994b; Maurer 1994; Heikkinen and Högmander 1994; Högmander and Møller 1995; Nowell and Jackson 1996). However, the commonest remains to determine the number of equal-area units of a grid system in which a species is found, with the size of the units for which occupancy is recorded tending to be positively correlated with the total area of the region over which the distributions are mapped (Fig. 3.2; Donald and Fuller 1998). Overall, there is a major disparity in the estimation of range sizes at the present time between the sophistication of the techniques that are available and the simplicity of those that are generally applied. Much of the work that has gone into deriving methods of estimation has seldom, if ever, been implemented, largely because of the crudity of the distribution data that typically are available.

Quinn *et al.* (1996) have documented significant, positive correlations across species between several different measures of range size (including latitudinal extent, 95 per cent latitudinal extent, longitudinal extent, 95 per cent longitudinal extent, latitudinal × longitudinal range, minimum convex polygon, number of $10 \times 10\,km$ squares occupied, number of $100 \times 100\,km$ squares occupied), for butterflies and freshwater molluscs across Britain (i.e. a partial analysis), based on data on occurrences at a resolution of $10 \times 10\,km$. For other assemblages, significant, positive correlations have also been documented between two or more such measures, particularly between latitudinal and longitudinal extent whose relationship gives some indication of the shape of geographic ranges (Lutz 1922; McAllister *et al.* 1986; Brown and Maurer 1989; Anderson and Marcus 1992; Pfrender *et al.* 1998; Murray and Dickman 2000)[1]. The different measures differ substantially in how robust they are to sampling effort, the level of which doubtless constitutes the greatest constraint on determining the real size of species geographic ranges (Gaston *et al.* 1996).

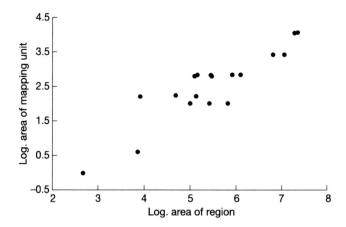

Fig. 3.2 The relationship between the sizes of the units for which the occupancy of species has been determined (km^2) and the area of the region over which occupancy is mapped (km^2). Based on data drawn from a number of published and unpublished sources.

3.3 Which range?

In addition to considerations of how to quantify the areas of geographic ranges, consideration of their size also highlights a need to decide *which* ranges are of primary interest. Using a rather idealized scheme, modified from that of Udvardy (1969), one can recognize several different kinds of geographic ranges. The *potential range* will not, in the main, concern us here, although it was touched on in the previous chapter and will receive some further consideration in a later section. It is the area over which a species would be found were all limitations to its dispersal to be overcome. The *realized range* is the smaller area within which the species does occur, and it has a number of components. It can be divided into a breeding area and a non-reproductive area. In the case of very many species of organisms, the breeding and non-reproductive areas are identical. For some species the difference between these two can, however, be extremely large. Thus, many seabirds forage over thousands of square kilometres of ocean, but breed on islands of just a few kilometres in extent. The great shearwater *Puffinus gravis*, for example, forages over much of the North and South Atlantic but is known only to breed on the islands of Nightingale, Inaccessible, Gough (all three in the Tristan da Cunha group), and Kidney (Falkland Is); the populations in the Tristan da Cunha group are thought to number several million breeding pairs (del Hoyo *et al.* 1992). In most landbird species the breeding range is larger than the non-reproductive, or wintering range, although Newton (1995) showed that for 62 bird species populations which breed in Eurasia and winter entirely in Africa there is a clear positive relationship between the size of the two ranges.

The breeding area of the realized range can be divided into a regular breeding area and a transient breeding area, the latter being unsuitable for prolonged occupation for breeding but being suitable for short periods. Transient breeding areas seem to occur commonly at the limits of species' ranges (Fig. 2.1), being occupied for breeding in years of favourable environmental conditions. Likewise, the non-reproductive area can be divided into one that is periodically but regularly inhabited for non-breeding purposes (e.g. wintering area of migrant species), and one that is unsuitable for prolonged occupation but in which the species may be found regularly for short periods for non-breeding purposes. For example, a number of bird species that breed and typically over-winter at high northern latitudes in the face of severe weather conditions periodically over-winter in considerable numbers much further south, exhibiting population eruptions.

As outlined in the opening chapter (Section 1.3), this book is predominantly concerned with the entire geographic ranges of species. A number of studies have documented interspecific relationships between the ranges of species measured over regions of different extent (i.e. at least for smaller extents covering only parts of the ranges of many if not all species; Hodgson 1986, 1993; Rapoport *et al.* 1986; Gaston and Lawton 1990; Jablonski and Valentine 1990; Daniels *et al.* 1991; Gregory and Blackburn 1998)[2]. Whilst commonly positive, the correlations are, however, typically rather weak (and in other analyses no significant relationships have been documented; e.g. Hughes 2000). On the basis of such relationships it is impossible to predict with any accuracy the size of the range of a species over a larger area from its range size over a smaller one; this makes sense, given the potentially substantial differences in the proportion of occupiable habitat over different-sized areas.

3.4 Species–range size distributions

Variation in the sizes of the geographic ranges of species in a taxonomic assemblage can effectively be summarized in terms of a frequency distribution. Such species–range size distributions tend to be unimodal with a strong right or positive skew. That is, most species have relatively small range sizes whilst a few have relatively large ones. This observation has been made for many extant assemblages (Table 3.3; Figs. 3.3, 3.4, 3.5, 3.6). It is, for example, illustrated by the Anseriformes, the wildfowl (Fig. 3.5). Species with relatively small breeding range sizes, such as the white-headed flightless steamer-duck *Tachyeres leucocephalus*, brown teal *Anas chlorotis*, red-breasted goose *Branta ruficollis* and maned duck *Chenonetta jubata* are the norm, whilst relatively widespread species, such as the northern pintail *Anas acuta* and mallard *Anas platyrhynchos* are the exception. Likewise, estimates of the

Table 3.3 Studies of species–range size distributions, for geographic range sizes; E, extent of occurrence; A, area of occupancy

Taxon	Region	Measure	Source
Plants			
seaweeds	Western Atlantic	E	Pielou (1977a)
plants	Mediterranean	A	Greuter (1991)
plants	Neotropics	A	Andersen *et al.* (1997)
Invertebrates			
sea urchins	Global	E	Emlet (1995)
prosobranch gastropods	Eastern temperate Pacific	E	Russell and Lindberg (1988a)
molluscs	Eastern Atlantic	E	Roy *et al.* (1995)
freshwater mussels	Atlantic and Eastern Gulf of Mexico coastal plains	A	Sepkoski and Rex (1974)
Evarthrus beetles	North America	E	Freitag (1969)
bombardier beetles	North America	E	Rapoport (1982)
Harpalus beetles	North America	E	Gaston (1998)
Nebria beetles	Nearctic	E	Gaston (1994a)
trogid beetles	Africa	A	Gaston and Chown (1999a)

(continued)

Table 3.3 *(continued)*

Taxon	Region	Measure	Source
bumblebees	Global	A	Gaston (1996c)
Fish			
freshwater fishes	North America	E	Anderson (1984b, 1985)
freshwater fishes	North America	E/A	McAllister *et al.* (1986)
freshwater fishes	West Africa	A	Hugueny (1990)
coral reef fishes	Global	E	McAllister *et al.* (1994)
Amphibians and reptiles			
amphibians	North America	E	Anderson (1984b)
amphibians	Australia	E	Anderson and Marcus (1992)
reptiles	North America	E	Anderson (1984b, 1985)
reptiles	Australia	E	Anderson and Marcus (1992)
turtles	Global	E	Hecnar (1999)
Birds			
birds	Australia	A	Schoener (1987), Gaston (1998)
non-passerines	Africa	A	Pomeroy and Ssekabiira (1990)
passerines	Africa	A	Pomeroy and Ssekabiira (1990)
birds	Australia	E	Anderson and Marcus (1992)
passerines	North America	E	Mehlman (1994)
birds	North America	E	Brown (1995)
procellariiforms	Global	A	Gaston and Chown (1999a)
wildfowl	Global	A	Gaston and Blackburn (1996b), Gaston (1998)
birds	New World	A	Blackburn and Gaston (1996a)
woodpeckers	Global	A	Blackburn *et al.* (1998a)
Mammals			
mammals	North America	E	Rapoport (1982), Anderson (1977, 1984a), Pagel *et al.* (1991)
mammals	Australia	E	Anderson and Marcus (1992), Smith *et al.* (1994)
Peromyscus mice	North America	E	Gaston (1994a)
mammals	Mexico	E	Ceballos *et al.* (1998)
mammals	Palaearctic	E	Letcher and Harvey (1994)
cetaceans	Global	A	Gaston and Chown (1999a)
primates	Africa	A	Eeley and Foley (1999), Eeley and Lawes (1999)
primates	Global	E	Gaston (1998)

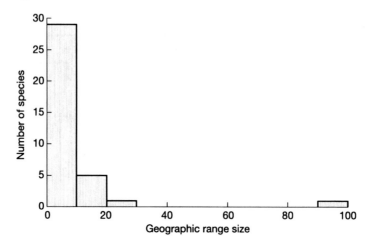

Fig. 3.3 Species–range size distribution for cats (Felidae). Range size × 10⁶ km². From data in Nowell and Jackson (1996).

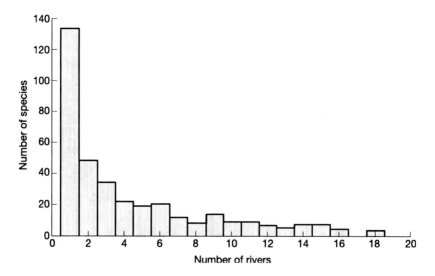

Fig. 3.4 Species–range size distribution for west African freshwater fishes, with range size measured as number of rivers. From Hugueny (1990).

present range sizes of the 36 extant species of cats (Felidae) indicate that the most widely distributed, the wildcat *Felis silvestris*, has a geographic range 680 times that of the most narrowly distributed, the Bornean bay cat *Catopuma badia* (Nowell and Jackson 1996). However, nearly two-thirds of cat species have range sizes that are less than 20 per cent of that of the most widespread species (Fig. 3.3).

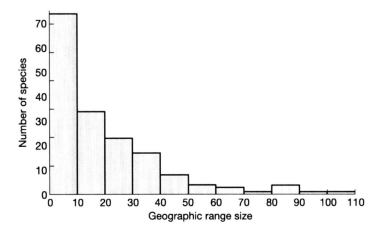

Fig. 3.5 Species–range size distribution for wildfowl, for breeding ranges. Range size is the number of 611 000 km² grid squares occupied. From Webb *et al.* (2001).

As well as many natural assemblages, right-skewed species–range size distributions are also exhibited for terrestrial assemblages of introduced species, including agricultural crops (Halloy 1999).

Published examples of the shapes of species–range size distributions for marine assemblages are scarce (but see Jablonski and Valentine 1990; McAllister *et al.* 1994; Gaston and Chown 1999a). Nonetheless, it seems almost certain that a markedly right-skewed pattern pertains here as for terrestrial assemblages. Thus, for example, McAllister *et al.* (1994) find that in a sample of 950 species of coral reef fishes 59 per cent have range spans of less than 6660 km, 33 per cent of less than 2220 km and 22 per cent of less than 1110 km (the Great Barrier Reef is approximately 2300 km long). Likewise, Roy *et al.* (1995) observe that the latitudinal ranges of extant species of eastern Pacific molluscs are commonly rather narrow, although the distribution is flatter and is not as markedly skewed as are many others (Fig. 3.6).

A similar pattern to the frequency of species range sizes observed for extant assemblages has also been documented for paleontological ones (e.g. Jablonski 1986a, 1987; Roy 1994). For example Jablonski (1987) found that most species of Late Cretaceous molluscs of the Gulf and Atlantic coastal plain of North America had relatively small geographic ranges. In short, the overall pattern seems long to have been a persistent characteristic of life on Earth. In fact, in virtually all published species–range size distributions in which range size is statistically untransformed, the most left-hand range size class is also the modal class. This pattern occurs almost irrespective of the particular method used to estimate or express the range sizes of the species.

Although there have been attempts to fit a variety of statistical model to species–range size distributions (e.g. Anderson 1977, 1984a, 1984b, 1985;

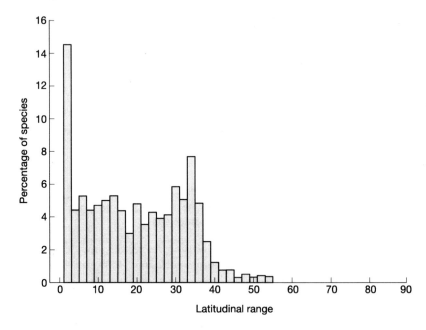

Fig. 3.6   Species–range size distribution for eastern Pacific molluscs, with range size measured as latitudinal extent (°). From Roy *et al.* (1995).

Gaston 1998; Buzas and Culver 1999), in general they tend toward an approximately normal distribution when geographic range sizes are subject to a logarithmic transformation (e.g. Anderson 1984a, 1984b; McAllister *et al.* 1986; Pagel *et al.* 1991; Anderson and Marcus 1992; Blackburn and Gaston 1996a; Hughes *et al.* 1996)[3]. However, they also appear consistently to acquire a mild to moderate left skew under such a transformation (Gaston 1998). This can be interpreted in one of two ways, either as inferring an excess or a lack of rare species relative to expectation under a lognormal model (for discussion in the context of species–abundance distributions, see Gregory 1994; Harte *et al.* 1999; Gaston and Blackburn 2000). There is some evidence that such distributions can effectively be approximately normalized by recognizing that range sizes have both a lower (at zero) and an upper bound (the limit to available space, e.g. the extent of a continent), and thus making use of folded rather than power transformations (the logarithmic transformation is a power transformation). A logit transformation seems appropriate, although this needs more exploration at the scale of entire geographic ranges rather than range sizes within a more constrained region (i.e. comprehensive analyses rather than partial ones; Williamson and Gaston 1999). This is important because patterns in the distribution of range sizes appear to differ systematically with spatial scale (Gaston 1994a; Gaston and Blackburn 2000).

The generality of the skewed distribution of the geographic ranges of species suggests some very general mechanistic interpretation. Whatever the fine details of its shape, a species–range size distribution based on the full geographic extents of species ranges must ultimately be a product of three processes: speciation, extinction, and the temporal dynamics of the range sizes of species between speciation and extinction.

3.4.1 Speciation

Speciation generates new species and hence additional geographic ranges, adding to a species–range size distribution. Its influence on such a distribution is, of course, more complex, and depends fundamentally on the geographic mode of speciation (allopatric, sympatric, etc., see Brooks and McLennan 1991), the form of the speciation event (or speciation model—anagenetic, cladogenetic, etc., see Wagner and Erwin 1995) and, to some extent, the interaction between the two (e.g. Archibald 1993; Wagner and Erwin 1995). However, consideration of the relationship between likelihood of speciation and the size of a geographic range has focused almost entirely on allopatric speciation, although other forms may perhaps be more widespread and frequent than has been acknowledged[4]. Moreover, emphasis tends to be laid on cases of complete geographic isolation, although this need not be a necessary precondition for speciation (Gavrilets *et al.* 1998).

The idea that species with larger geographic range sizes have a greater probability of allopatric speciation dates at least to Darwin (1859) and continues to attract support (e.g. Terborgh 1973; Marzluff and Dial 1991; Budd and Coates 1992; Roy 1994; Wagner and Erwin 1995; Tokeshi 1996), apparently on the premise that larger geographic ranges present more opportunities for subdivision. Rosenzweig (1975, 1978, 1995) argues that on a purely probabilistic basis species with larger geographic range sizes are more likely to undergo speciation, because the chance of their ranges being bisected by a barrier is greater than for a small range size. Differentiating, as Rosenzweig does, between two kinds of barriers, 'knives' which have beginnings and ends (serving to sever a range into two parts), and 'moats' which surround their isolates, then strictly this assertion is only true of moats. Large geographic ranges will tend to engulf knives, such that they do not engender speciation, and the probability of division will have a peak at intermediate range sizes. Rosenzweig (1995) maintains that this is unlikely to occur because there are no, or virtually no, species with geographic ranges so large that reducing them would make them an easier target for barriers. However, whether this is so depends critically on the frequency distribution of barrier sizes. If most barriers are small to intermediate in size, relative to the range sizes of widespread species, then intermediate-sized ranges may indeed have a higher probability of speciation. Given that this seems likely to be true for most kinds of barriers that one can think of (most rivers, mountain chains, etc. are reasonably short compared with the extents of occurrence of many species, and only a few are very long), then species with intermediate-sized ranges will give rise to more descendants (see Chown 1997; Gaston 1998;

Gaston and Chown 1999a; Chown and Gaston 2000). Such an effect would be enhanced because barriers seem far more likely, by chance if nothing else, to take the form of knives than of moats.

There are two additional reasons for believing that the likelihood of speciation and geographic range size are linked by a peaked function. First, widespread and abundant taxa may often possess well-developed dispersal abilities (which may perhaps have enabled them to become widespread in the first place; but see Section 3.5.8) and should, as a consequence, have a strong proclivity to maintain gene flow amongst populations, which will tend to inhibit speciation (e.g. Mayr 1963, 1988; Stanley 1979; Flessa and Thomas 1985; Lieberman *et al.* 1993; Garcia-Ramos and Kirkpatrick 1997). Narrowly distributed and locally rare taxa with poor dispersal abilities (and patchy populations which may tend to form isolates) will tend to have higher speciation rates. Although the extent of connectivity of the ranges of very widespread species is, in probably most cases, reasonably limited (without human assistance, individuals do not tend to flow freely between continents or ocean basins; see James *et al.* 1999), recent models have demonstrated that both enhanced gene flow and weak environmental gradients are liable to promote genetic cohesion (Rice and Hostert 1993; Kirkpatrick and Barton 1997).

The second additional reason for expecting likelihood of speciation and range size to be related according to a peaked function is that there is a, generally, positive interspecific relationship between population size and range size (discussed at greater length in Section 3.5.9). This relationship is such that population size increases at a disproportionately faster rate with increasing range size, and hence local density also increases with increasing range size (Brown 1984, 1995; Gaston 1996d; Gaston *et al.* 1997a). In consequence, density-dependent dispersal probability would tend disproportionately to increase the cohesion of ranges with increases in their size. Not only should higher densities result in higher numbers of individuals moving between disjunct populations on a purely probabilistic basis, but gene flow from populations at a range centre can act as a powerful inhibitor of change in more peripheral populations, thus precluding speciation (Section 2.4; Hansen 1978, 1980; Jabonski and Lutz 1983; Garcia-Ramos and Kirkpatrick 1997; Gavrilets *et al.* 1998). Gavrilets *et al.* (2000), performing large-scale individual-based simulations, find that smaller range sizes should have higher speciation rates, because they are characterized by reduced dispersal ability and lower local densities.

These assertions assume that all else is equal. Of course, many factors influence probabilities of speciation (perhaps including both species and environmental factors—generation time, body size, temperature, etc.), and it is unclear how significant range size might be relative to others within even a reasonably homogenous taxonomic group. Moreover, the above observations refer specifically to instantaneous rates of speciation. The relationship between the probability of speciation and geographic range size may be rather different if one considers this probability over the lifetime of an ancestral species, assuming ancestral persistence after having given rise to one or more daughter species (a possibility that many evolutionary biologists find inconvenient to contemplate!). If species with large geographic ranges persist for longer periods, then they may leave more descendants than species with smaller ranges. In analyses of two Neogene clades of Foraminifera and an Ordovician family of gastropods, species with larger ranges were likely to leave more descendants in two cases, but not in the third, and in all three cases species that persisted for longer were likely to leave more descendants (Wagner and Erwin 1995).

The impact of speciation on the form of species–range size distributions will rest not only on the frequency with which geographic ranges of different size speciate, but also on the pattern of sub-division of the ancestral range. If it is ancestral species with relatively small geographic range sizes which are most likely to speciate, then the products of any range division can only be two small ranges (be they two daughters, or a daughter and its ancestor if the latter persists). If species with large geographic range sizes are more likely to speciate then, depending on the asymmetry of division, the outcome may span one large and one small range size through to two ranges each half the size of that of the ancestor at speciation. If we continue to regard patterns of speciation in terms of simple random events then, even with species with small range sizes being the more likely to speciate, the most likely immediate products of speciation are two species one of which is more widely distributed than is the other. A perfect fifty–fifty split is highly improbable.

The question of the degree of asymmetry in the range sizes of sister-groups, immediately postspeciation, is closely related to the issue of whether allopatric speciation is best typified by a peripheral isolation model or by a vicariance model. In the former, peripheral isolates form by waif dispersal (establishment of a new population through long-distance movement across a barrier), microvicariance (physical division of a previously continuous distribution), or range retraction (causing peripheral populations to become isolated; Frey 1993). Here, the relatively widespread ancestral species is likely to change little whilst the peripheral isolate diverges (Glazier 1987). In the vicariance model, a subdivision of the range of an ancestral species occurs, such that there is cessation of contact between the two subpopulations, giving rise to new species. Here, both daughters of the ancestral range are likely to diverge, and the ancestral species will cease to exist. The two models are plainly very closely related, and may in some sense be seen as constituting points on a continuum of speciation processes. However, vicariant speciation may potentially result in any degree of asymmetry in the initial range sizes of daughter species, and is often portrayed as generating very similar sized ranges. In contrast, peripheral isolation results, immediately postspeciation, in a highly asymmetrical split.

Both peripheral isolation (e.g. Kavanaugh 1979a; Ripley and Beehler 1990; Levin 1993; Chesser and Zink 1994) and vicariance models (e.g. Cracraft 1982, 1986; Cracraft and Prum 1988; Lynch 1989) have significant support. The relative frequency of the two modes of speciation remains a point of some contention (e.g. Bush 1975; Lynch 1989; Ripley and Beehler 1990; Brooks and McLennan 1993; Frey 1993; Chesser and Zink 1994)[5]. Resolution of the issue rests in major part on the extent of postspeciational change in geographic range sizes. This will, to a marked degree, determine the extent to which the present-day distributions of species can be used to reconstruct past patterns of speciation (Lynch 1989; Brooks and McLennan 1991).

If dispersal is important, then it may entirely obscure the relative positions of, say, sister-taxa at the time of their divergence. If it is not important then this will not be a problem. It is, of course, important, and it is increasingly clear that species originating through different modes of speciation may display similar present-day patterns of distribution (García-Ramos *et al.* 2000).

3.4.2 Range dynamics

Regardless of its area at speciation, the geographic range size of a species will invariably change through the duration of its persistence, perhaps in complex ways, as a result of variation in the constraining factors (now commonly associated with human activities). Elsewhere I have, for convenience, termed the temporal dynamics of ranges between speciation and extinction as 'range transformations' (Gaston 1998; Gaston and Chown 1999a). However, in the present context this results in a confusing multitude of uses of the term 'transformation', and I will continue to use 'dynamics', whilst acknowledging that both speciation and extinction are dynamic events with regard to geographic range size.

There is a potentially infinite set of trajectories that the geographic range sizes of species could take between speciation and ultimate extinction. However, just as Lawton (1992) argued that it is unlikely that there are 10 million kinds of population dynamics (on the assumption that there are 10 million species of plants and animals), but rather a multitude of essentially trivial variations on a few common themes, so there seem likely to be a few common themes in range size dynamics. The number of such themes remains unclear, but here I will consider five different broad classes of dynamics, although these are not necessarily of equivalent status (Fig. 3.7; see also Gaston and Blackburn 1997a; Gaston and Kunin 1997; Gaston 1998; Gaston and Chown 1999a).

1. *Stasis* (Fig. 3.7a and c)—this is an unrealistic model. If the geographic range sizes of species remained essentially constant between speciation and extinction, then one would expect there to have been a very marked decline in the average range sizes of species over evolutionary time, and an absence of widespread species, as speciation progressively fragmented ancestral ranges. Importantly, consideration of this model begs the general question of what the trends in the mean, maximum, and minimum geographic range sizes across species in particular higher taxa actually have been over long periods of time. I am aware of few data on this subject, although it seems likely that there have been systematic shifts, particularly associated with the break-up of the continental land masses.

2. *Stasis postexpansion* (Fig. 3.7b)—in this model, the geographic ranges of species increase in size rapidly after speciation, and the resultant area is then approximately maintained until shortly before extinction (Jablonski 1987; see also Rosen 1988). This model requires that species spread very rapidly postspeciation. An explicit test of this model is difficult, but Jablonski (1987) derived support from the observation that the range sizes of molluscs arising in the 2 million years

before the end-Cretaceous mass extinction (and whose geological durations were truncated by this event) were indistinguishable from the range sizes of molluscs arising in the previous 14 million years.

3. *Age and area* (Fig. 3.7d and e)—Willis (1922) argued that the geographic ranges of species followed a trajectory of steadily increasing size the longer they persisted, presumably culminating in a rapid decline to extinction or disappearance through cladogenesis. This hypothesis underwent vigorous debate, was widely rejected (e.g. Gleason 1924; Stebbins and Major 1965; Stebbins 1978), and has been described by Brown (1995, p.102) as a 'quaint anachronism'. In its simplest form it cannot be correct, because there are ample examples of young, widely distributed species and old species which are very restricted in their distribution (although species on the verge of extinction must have small geographic ranges). Nonetheless, it continues to be maintained in some circles that there is a broad, positive interspecific correlation between age and area (although such a pattern

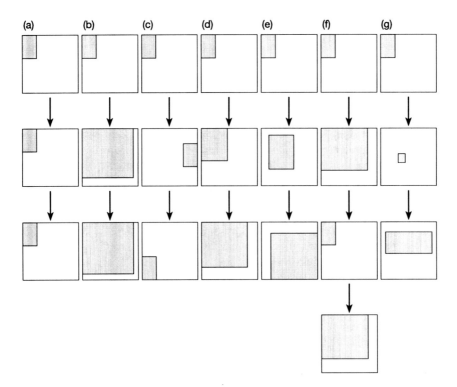

Fig. 3.7 Potential ways in which geographic range size and position can vary through time: (a) stasis, (b) stasis post-expansion, (c) stasis with positional shift, (d) age and area, (e) age and area with positional shift, (f) cyclic, and (g) idiosyncratic. In each case the shaded box represents the geographic range and the arrows the transitions between different time periods. From Gaston and Chown (1999a).

need not follow from an intraspecific age and area relationship), albeit not necessarily of strong predictive value (McLaughlin 1992), and some recent analyses have reported such a pattern, at least amongst small numbers of closely related species (Taylor and Gotelli 1994) and at the generic level (Miller 1997).

4. *Cyclic* (Fig. 3.7f)—various authors have regarded the range size dynamics of species as essentially cyclic (e.g. Dillon 1966), and this is most explicitly expressed in the concept of the taxon cycle (Wilson 1961; Ricklefs and Cox 1972, 1978). In the taxon cycle species pass through a sequence of range expansion, evolutionary differentiation between populations, and local extinction creating gaps within the distribution, with taxa either becoming globally extinct or undergoing a fresh phase of range expansion. Evidence has been produced both in support of (Ricklefs and Cox 1972; Glazier 1980; Rummel and Roughgarden 1985; Roughgarden and Pacala 1989; Ricklefs and Bermingham 1999) and against (Pregill and Olson 1981; Liebherr and Hajek 1990; Losos 1992) the existence of taxon cycles for particular assemblages; the observed pattern is likely to be determined by the geographic pattern of isolation. Exceptions to a taxon cycle concept of range contraction and expansion may include range size changes associated with contraction of species ranges to refugia and their subsequent expansion (see e.g. Kavanaugh 1979b; Lynch 1988), and the processes that have given rise to bipolar/amphitropical distributions (e.g. Crowson 1981; Crame 1993). However, the nature of refugia and their geographic location, and the processes leading to bipolar distributions, remain polemical (Lynch 1988; Crame 1993; Bush 1994).

5. *Idiosyncratic* (Fig. 3.7g)—there need, of course, be no general pattern of change in the geographic range sizes of species between speciation and extinction. Different species may exhibit entirely idiosyncratic trajectories. This would be in keeping with the continual adaptation and change that may result from responding to the demands of the Red Queen (Van Valen 1973a; Ricklefs and Latham 1992; Vrba 1993), and with the changes in the distributions of some species over the past few decades (e.g. Frey 1992; Burton 1995; Parmesan 1996). If such a pattern prevails, then one might expect to see little similarity in the geographic range sizes of closely related species (see Section 3.5.1), unless the traits which influence range size are strongly phylogenetically conserved, in which case in climatically and ecologically similar regions the distributions of close relatives might be expected to fluctuate in parallel (Ricklefs and Latham 1992).

Some of these models have important implications. Thus, for example, if the phase of increase in range size is brief relative to the phase of decline, then over any given evolutionary time period most species will be observed to be declining in their geographic range sizes, even in the absence of any over-arching directional pressure on range sizes. This would obviously be significant when considering the consequences of such a dynamic.

In practice, the temporal trajectories of the geographic range sizes of species are exceedingly difficult to determine over long periods of time. In principle, suitable data might be forthcoming from the fossil record, but whether these could be obtained at a sufficient temporal resolution, and sufficiently free of confounding influences, is extremely doubtful. Evidence in support of different models, to date, has arisen largely from comparison of the geographic ranges of species estimated to be of different ages. This assumes that different species follow broadly similar temporal trajectories in their range sizes and that any observed differences are, in major part, a consequence of differences in their ages. Such an assumption seems most likely to be true of groups of closely related species. Even then it is problematic.

Using data for several monophyletic groups of birds with well resolved phylogenies which included all or the vast majority of extant species, Webb and Gaston (2000) examined interspecific relationships between current geographic range size and evolutionary age (approximate time since divergence). They found that range sizes appear to expand relatively rapidly post speciation, at least on evolutionary time scales; subsequently, and perhaps more gradually, they then decline as species age (Fig. 3.8; see also Flessa and Thomas 1985). The capacity of ranges to expand quickly is attested by studies of the spread of species when they have been introduced into regions in which they do not occur naturally. Recorded rates of spread can be very high (Figure 3.9; e.g. Wing 1943; Lynch 1989; Hengeveld 1989; Grosholz 1996; Veit and Lewis 1996; Williamson 1996). The horse was reintroduced to North America (having previously been exterminated, whether by humans or some other factor remains a topic of hot debate; Flannery 2001; Grayson 2001) by Europeans and occupied much of the continent within 30 years, although this area was subsequently much reduced (Flannery 2001). Mack (1997, p.210) observes that 'a striking feature of post-Columbian terrestrial plant invasions has been the swiftness with which these events have unfolded', with a 'lag phase' following their arrival, when they comprise a small and inconspicuous population of founders (which will be a permanent state for many invaders), being followed by quick expansion such that the majority of the new range is occupied in less than the whole lag phase. The lag before a rapid expansion appears to be very common. It may be associated with the ability of species only to spread locally when densities are low, but when numbers have built up successful long-distance colonizations may become more likely (if only for reasons of numbers of possible colonists; cf. Veit 2000) and spread then rapid (Cousens and Mortimer 1995); range expansion often proceeds through the establishment of outlying populations and subsequent 'back-filling' of unoccupied area (Ibrahim *et al.* 1996). Equally, the lag might be associated with the time required for selection for long distance dispersers to operate (Section 2.2.1). Whatever, the spread of a number of species is well modelled as an exponential increase (giving a linear relationship between log. area occupied and time), implying that spread is not a case of simple, constant, radial expansion (which would give a linear relationship for the square root of area occupied; Cousens and Mortimer 1995). A fascinating feature of the interspecific relationships between current geographic range size and evolutionary age of Webb and Gaston (2000), is that species with geographic ranges that have been seriously impacted by human activities tend to constitute conspicuous outliers to any general patterns, with markedly smaller

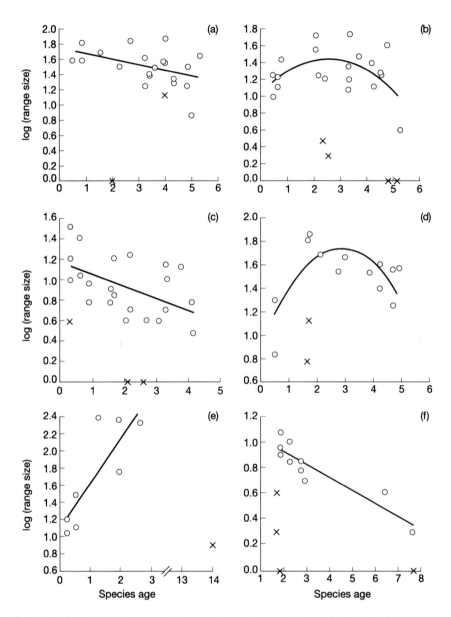

Fig. 3.8 Plots of global geographic range size against species age (Myr) for (a) Old World *Acrocephalus* and *Hippolais* reed warblers, (b) New World *Dendroica* wood warblers, (c) New World *Icterus* orioles, (d) storks, (e) gannets and boobies, and (f) albatrosses. Crosses represent species excluded from regression analyses, island endemics and species that have suffered substantial recent declines in range sizes due to anthropogenic factors. From Webb and Gaston (2000).

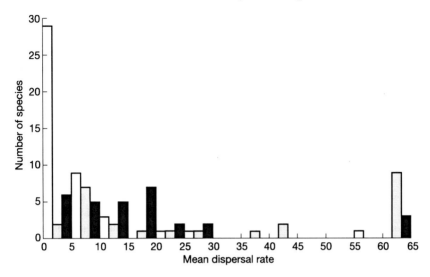

Fig. 3.9 Mean radial dispersal rates (km year^{-1}) for a diversity of species of plants (open bars), invertebrates (grey bars) and vertebrates (solid bars). From data compiled by Rapoport (2000).

range sizes than would otherwise have been expected for their ages. As Marren (1999) observes, rare species can be divided into three Shakespearean categories, some were born rare, some achieved rarity, and others had rarity thrust upon them.

Whatever the overall life-time trajectory of the geographic range size of a species, if any, it is clear that there will be a great deal of variance about this trend. The interesting issue then becomes what constitutes trend and what constitutes variance. Much of the natural variation in the range sizes of species that we observe over ecological time doubtless contributes such variance. Gaston and He (2002) have built on this argument in developing a stochastic differential equation to describe the dynamics of the range size of an individual species. Based on a power relationship between abundance and range size (Section 3.5.9), they derive a limiting stationary probability model to quantify the stochastic nature of the range size for that species at steady state (varying about some equilibrium), and then generalize this model (assuming that the range sizes of all species follow a similar stochastic process, and that carrying capacity is a random variable) to the species–range size distribution for an assemblage. The model fits well to several empirical data sets of the geographic range sizes of species in taxonomic assemblages, and may provide the simplest explanation of species–range size distributions to date, although it ignores any influence of speciation and extinction on the pattern of range size and makes a number of possibly contentious assumptions.

As mentioned earlier, the temporal dynamics of range sizes are not independent of speciation events. If there is ancestral persistence after speciation events then the geographic range sizes of ancestors will be reduced by those events (to what extent will rest on whether speciation occurs by peripheral isolation, which will have little impact, or vicariance, which may have a major impact). Likewise, if there is no ancestral persistence then the final range size of an ancestor will be that which it has at the time of speciation, when it becomes extinct.

3.4.3 Extinction

The general interaction between the mode and form of speciation and geographic range size determines that range size with which a species is effectively 'born' and perhaps subsequent alterations to its range size during its existence if the species gives rise to other species (if speciation can be accompanied by ancestral persistence). The interaction between extinction and geographic range size determines the length of that existence. There is no strong body of theory relating risk of extinction and range size, such as exists for the relationship between risk of extinction and population size (e.g. Lande 1993, 1998; Lynch and Lande 1998). However, persistence would seem to increase for larger geographic ranges, on the grounds, first, that such ranges typically comprise larger numbers of individuals (Section 3.5.9), and thus have a smaller chance of a random walk to extinction, and, second, that larger ranges spread the risk of adverse conditions affecting distributions as a whole (it is less likely that all areas will be in decline or undergo local extinction simultaneously). Thus species with larger geographic range sizes are more resistant to global extinction, and persist for longer, than do those with smaller range sizes (an observation which dates at least from Lamarck; see McKinney 1997).

There appears to be reasonable empirical evidence for a positive relationship between measures of time to extinction and geographic range size for various paleontological species assemblages (Jackson 1974; Hansen 1978, 1980; Koch 1980; Jablonski et al. 1985; Jablonski 1986a, b, 1987; Buzas and Culver 1991—but see also Stanley 1986; Stanley et al. 1988; for analyses above the species level, see Westrop 1989, 1991; Jablonski and Raup 1995; Erwin 1996; Flessa and Jablonski 1996; Smith and Jeffery 1998). It is not always especially strong and may break down during periods of mass extinction (Jablonski 1986a; Norris 1991). A component of this pattern is doubtless an artefact resulting because more widely distributed species tend also to be locally more abundant (Section 3.5.9; the pattern has been documented for paleontological assemblages as well as contemporary ones; Buzas et al. 1982; McKinney 1997; but see Koch 1980) and thus are more likely to be recorded in the fossil record (Russell and Lindberg 1988a, 1988b); persistence will, of course, always tend to be underestimated to some extent, regardless of abundance. However, this effect seems insufficient to explain the observed pattern in its entirety (Jablonski 1988; Smith 1994), although it remains a moot point how much of the variation in observed relationships between persistence and range size can be explained as a product of sampling (a number of techniques have been developed to improve the reliability of estimated stratigraphic durations; e.g. Marshall 1991, 1994; Benton 1994).

Positive relationships between geographic range size and persistence based on paleontological data may also arise artefactually if insufficient attention is paid to the method of determination of range size (a challenging problem for extant species, without the additional complications presented by the fossil record). What is of interest is the relationship for instantaneous measures or at least those integrated only over a short period. However, by definition these are impossible to obtain from a fossil record that, by its very nature, synthesizes information over, often long, time periods. Where studies are explicit about the way in which range size is calculated, it appears that in some cases this is done by summing the spatial extent of localities (often number of provinces) over the total geological duration of the species, or the geological period under investigation (e.g. Westrop and Lundvigsen 1987; Stanley *et al.* 1988; see also Koch 1987; Koch and Morgan 1988; Budd and Coates 1992; Roy 1994). Such an 'allochronic extent of occurrence' estimate (Gaston and Chown 1999a) would, of course, only provide a reliable indication of the geographic range size of a species if it could be demonstrated that subsequent to speciation, species rapidly attain the range size that will characterize them for the bulk of their geological duration, or that a species range size remains static throughout its duration (which as already observed seems unlikely).

As well as paleontological species assemblages, evidence for a negative relationship between geographic range size and likelihood of extinction is often drawn from largely extant ones (e.g. Laurila and Järvinen 1989; Mace 1994; Gaston and Blackburn 1996b; Purvis *et al.* 2000; Strayer 2001). Where analyses have concerned global extinction, they have tended to be based on relationships between the sizes of the current or recent historic geographic ranges of species and the relative risk of extinction they are accorded under various schemes for classifying threat status. The obvious limitation of this latter approach is that it may be entirely circular, if range sizes are used in or influence the determination of threat status (as they often do, see below). However, where careful attempts have been made to control for such effects, the relationships have remained (Purvis *et al.* 2000). These relationships clearly, however, have much variance about them, with some species with very small range sizes apparently persisting for long periods, and some with large range sizes being at high risk of extinction.

The importance of range size *per se*, as against other factors, in determining risk of extinction remains unknown. Nonetheless, comparison of the estimated extinction rates of species in vertebrate groups in North America and the range sizes of species in those groups does suggest that the most restricted groups, freshwater fishes and amphibians, are also the most threatened (Table 3.4).

On balance, there is almost certainly a negative correlation between geographic range size and risk of extinction, but the evidence is poorer than is commonly supposed, and the relationship is doubtless weaker than seems frequently to be assumed; 'The argument that a species is in no danger because it is very common, is a complete fallacy ...' (Elton 1927). It seems highly likely that the relationship is also typically strongly non-linear, with species with small ranges having disproportionately higher rates of extinction than those with intermediate range sizes.

Table 3.4 The geometric mean range size (10^5 km²) of species in different vertebrate groups in North America, and estimates of the recent and predicted future species extinction rates (per cent loss per decade) of these groups; ranges of values are given where no overall figure was provided for a group, but separate figures for constituent subgroups; from data in Anderson (1984b) and Ricciardi and Rasmussen (1999)

	Mean range size	Extinction rate estimates	
		Recent	Future
Freshwater fishes	0.82	0.4	2.4
Amphibians	0.9–4.6	0.2	3.0
Reptiles	2.8–8	0	0.7
Birds	16	0.3	0.7
Mammals	5.8	0	0.7

The existence of a negative relationship between the likelihood of extinction of a species and the size of its geographic range has provided one of the foundations for many schemes designed to categorize extant species as at different risks of extinction in the immediate to near future. Thus, the new IUCN Red Data Book criteria (IUCN 1994) include as one route to classifying species as Critically endangered, Endangered, or Vulnerable, criteria based on combinations of absolute limits to geographic range size, and evidence of continuing decline, extreme fragmentation, limitation to a few independent subpopulations, or extreme fluctuations (for applications, and discussions, of these criteria see Mace *et al.* 1992; Collar *et al.* 1994; Pinchera *et al.* 1997; Hallingbäck 1998; Hallingbäck *et al.* 1998; IUCN/SSC Criteria Review Working Group 1999). Extents of occurrence or areas of occupancy, range-wide or over smaller areas, are also employed as indicators of risk of extinction in many other such schemes (e.g. Nowell and Jackson 1996; Kirchhofer 1997). As explicitly observed in the IUCN documentation, the inclusion of absolute limits to range size as criteria for listing species as under different degrees of threat begs questions of consistency in the way in which range sizes are being measured, particularly for areas of occupancy which depend so fundamentally on the spatial resolution at which they are determined. However, whilst the relationship between range size and likelihood of extinction is much more poorly understood than that between population size and likelihood of extinction, and is expected to be a much weaker relationship, the vastly greater availability of information on the distributions of species compared with their abundances has placed heavy reliance on measures of geographic range size as the basis of relative estimates of extinction risk.

It is also clear that human activities have the potential rapidly to bring even relatively widespread species to the brink of extinction and beyond. In the terrestrial realm the best example is doubtless that of the passenger pigeon

Ectopistes migratorius. When Europeans reached North America, this species is estimated to have had a population of perhaps 3 to 5 billion individuals (perhaps comparable to the current entire bird population of North America) with a notably easterly distribution across the continent (Schorger 1955). Nonetheless, forest destruction and fragmentation, facilitated by high levels of exploitation and disturbance, resulted in dramatic declines such that the species was scarce in the 1880s and very rare by the 1890s, with the last authentic recorded sightings occurring at the turn of the century and the last individual dying in Cincinnati Zoo in 1914 (Schorger 1955; Blockstein and Tordoff 1985; Bucher 1992).

Murray and Dickman (2000) found that species of Australian mammals that have become extinct in the last 200 years did not have smaller historical ranges than other pre-European species, again emphasizing the potential for rapid declines to be caused by human activities. In the marine realm, long-term surveys on the continental shelf between the Grand Banks of Newfoundland and southern New England have revealed that one of the largest skates in the north-west Atlantic, the barn-door skate *Raja laevis*, has declined dramatically over the past 45 years throughout its range and, if the trend continues '… could become the first well-documented example of extinction in a marine fish species' (Casey and Myers 1998).

In terms of the effect of extinction on the shape of species–range size distributions, any negative relationship between persistence and range size will tend differentially to remove ranges from the left-hand side of the distribution. As yet, we are not in a position to combine this effect, the relationship between range size and speciation, and the temporal dynamics of range size between speciation and extinction, to yield a single dynamic model of species–range size distributions. There are at this point too many unknowns, particularly over the most appropriate temporal dynamics to assume. By contrast, rather more progress has been made in understanding species–body size distributions, the frequency of species with different body sizes, which must also be generated by the combined effect of speciation, extinction, and variation in the trait of interest (Maurer *et al.* 1992; Johst and Brandl 1997; Gaston and Blackburn 2000). It is apparent that the observed shape of these distributions is most probably under small-biased speciation and large-biased extinction (Kozłowski and Gawelczyk 2002). However, the case is somewhat simpler in that body size variation during the life time of a species is trivial compared with that between species in an assemblage and can thus effectively be ignored (although systematic patterns of intraspecific size variation may occur both spatially and temporally), and speciation does not change the body size of the ancestral species. Regarding the shape of species–range size distributions, a best guess would be that the answer lies in a model that assumes a bias toward allopatric speciation in small to intermediate-sized ranges, with most such events resulting in highly asymmetric splitting of ancestral ranges, a broadly hump-shaped trajectory of range size between speciation and extinction, and a greater likelihood of extinction by species with small range sizes.

3.5 Patterns in range size variation

The evolutionary dynamics of geographic ranges, including speciation, extinction, and fluctuation in size between these two events, in major part represent the interplay of species characteristics and their abiotic and biotic environments. Those interactions result in the distributions of different species being limited by different sets of factors, with those factors often varying between different times and places. Despite this complexity, a number of patterns of systematic variation in range sizes between species have been proposed to exist. In the rest of this chapter these patterns will be considered.

3.5.1 Taxonomic group

There have been few formal comparisons of the sizes (mean, minimum, maximum, etc.) of the geographic ranges attained by species in different major taxonomic groups. A number of broad generalizations have, nonetheless, been made, but often with rather limited empirical support and little data on which to base appropriate tests. These include that:

1. On average, protists have larger geographic ranges than do metazoans (Fenchel 1993).

2. Freshwater plants have particularly large geographic ranges (Marren 1999).

3. On average, plants and insects have smaller geographic ranges than do vertebrates (Gaston 1994a).

4. The smallest ranges of vascular plants and fishes are several orders of magnitude smaller in size than the smallest ranges of birds and mammals (Brown *et al.* 1996).

5. Land snails have particularly small geographic range sizes (Solem 1984; see also Cameron 1998).

6. On average, birds have larger geographic ranges than do mammals (Anderson 1984a, 1984b; Anderson and Marcus 1992).

Almost all of these assertions seem to follow from the relative habitat availabilities, dispersal abilities, and home range requirements of the different groups. At the scale of Britain, Gaston *et al.* (1998a) determined that the median areas of occupancy of species in different taxa decreased in the order birds, mammals, butterflies, terrestrial molluscs, dragonflies, aquatic molluscs, vascular plants, moths, and liverworts (because of the right-skewed form of species–range size distributions, the median is a better measure of central tendency than the mean). How well this ranking generalizes to entire geographic ranges and other regions is not known, although it would seem to fit well with several of the above preconceptions, and some data (e.g. Rapoport 1982; Anderson 1984a; Anderson and Marcus 1992). Brooks *et al.* (2001) observe that in their data base of the distributions of species across Africa, birds have a median range size of 144 1° × 1° cells, mammals a median of 33, snakes 14, and amphibians 10. However, the degree of difference in the range sizes of birds is probably inflated by biases in the data.

Although there are doubtless important systematic differences in geographic range sizes between major groups of organisms, unlike many other biological variables (e.g. body size, life history traits) range size exhibits remarkably little phylogenetic conservatism within such a group (Arita 1993; Brown 1995; Gaston and Blackburn 1997b; Blackburn *et al.* 1998a; Gaston 1998; Webb *et al.* 2001; for similar results for range sizes at meso-scales see Hodgson 1993; Peat and Fitter 1994; Kelly and Woodward 1996; Böhning-Gaese and Oberrath 1999). This can be demonstrated by examining the proportion of variation in range sizes between species that is accounted for by different taxonomic levels in a hierar-chical nested analysis of variance (ANOVA). Thus, for example, for the wildfowl species of the world more than 80 per cent of variance in breeding range size is accounted for at the level of the species, with most of the rest at the level of the tribe (Fig. 3.10). This contrasts with only about 10 per cent of variation in body size

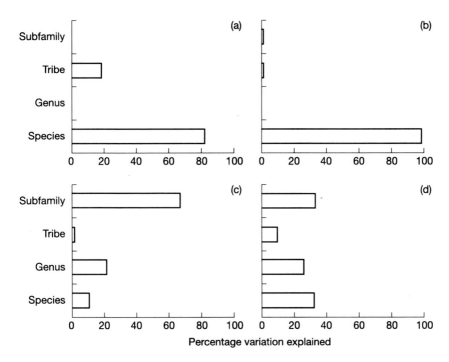

Fig. 3.10 The percentage of variation explained at different taxonomic levels in the wildfowl (Anseriformes), for (a) breeding geographic range size (number of 611 000 km² grid squares occupied), (b) maximum global population size, (c) body size (mean female body mass, g), and (d) average clutch size. Variance components are estimated using a nested ANOVA. From Webb *et al.* (2001).

being accounted for at the level of species, *c.*22 per cent at the level of genus, and *c.*67 per cent at the level of subfamily. The lack of phylogenetic conservatism is also evidenced using a bivariate plot of the range sizes of one species in different pairs of sister species against that of the other species. Figure 3.11 shows such a plot for wildfowl, with the axis on which either species of a given sister pair is plotted being randomized, and it is apparent that there is no relationship. Closely related species do not have similar geographic range sizes. In contrast, an identical plot, based on the body sizes of the same species pairs shows a strong positive relationship; closely related species tend to be quite similar in body size. The lack of phylogenetic conservatism in geographic range size conforms with common experience. For most groups of organisms it is easy to think of closely related species at least one of which is widespread and another restricted in its distribution. Given that many physiological parameters, such as temperature tolerance, show substantially greater phylogenetic constraint than range size, it again seems unlikely that in any simple sense climate is playing the predominant role in limiting observed range in many groups (Chapter 2).

Given the often substantial differences in the range size of species in different taxonomic groups it is impossible to state what the average range size of all extant species might be. Based on a very narrow range of taxa and with little attention to the distribution of species across the Earth, in estimating the global number of populations and their rate of loss, Hughes *et al.* (1997) use an average range size (extent of occurrence) for a species of 2.2 million km^2. This is an area about the size of Greenland or of Saudi Arabia, which seems somewhat on the large size. Of the taxonomic groups with at least moderate species richness, most is known about the

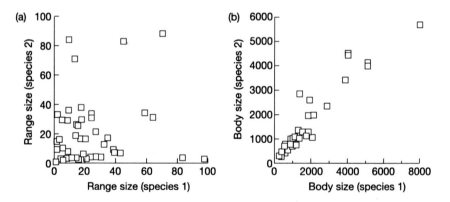

Fig. 3.11 Typical examples of comparisons of (a) breeding geographic range sizes (number of 611 000 km^2 grid squares occupied) and (b) body size (mean female body mass, g), in 46 pairs of sister species of wildfowl (Anseriformes). For 100 randomizations of the attribution of species 1 and species 2 of each pair to the *x* or *y* axis, there was no significant correlation for range sizes ($r_s = 0.180 \pm 0.002$ (SE), $n = 100$), but body sizes were highly correlated ($r_s = 0.940 \pm 0.001$). From Webb *et al.* (2001).

geographic ranges of birds. Stattersfield *et al.* (1998) judge that 28 per cent of all landbird species have had a breeding range size of 50 000 km^2 or less throughout historical times.

3.5.2 Terrestrial versus marine systems

It is widely held that marine species have larger average geographic range sizes than do terrestrial ones. This would make sense, particularly for pelagic and benthic marine species, since the oceans have a greater extent than the land masses (oceans cover 71 per cent of the Earth's surface and land 29 per cent), exhibit greater contiguity, are three-dimensional (averaging *c.* 3.8 km in depth and having an overall volume of 1370 × 10^6 km^3), and many terrestrial groups of organisms diversified on the continents after they had fragmented. However, the difference in range sizes may not be as marked as this might imply, as there is growing evidence that marine populations are probably not as open as has often been supposed and that there may commonly be barriers to gene flow, both horizontally and vertically (with water depth; e.g. Palumbi 1994; Bahri-Sfar *et al.* 2000; Barber *et al.* 2000; Bernardi 2000; Cowen *et al.* 2000; Terry *et al.* 2000).

Again, empirical analyses that demonstrate real differences in range sizes between species in the terrestrial and marine realms are largely wanting. They are also complicated by phylogenetic considerations, given the relatively small number of groups that have at least moderately closely related marine and terrestrial representatives. However, Rapoport (1994) reports that marine mammalian carnivores have a greater level of 'cosmopolitanism' than terrestrial mammalian carnivores.

It has long been held that, in general, marine species are at less risk of global extinction than terrestrial species, in part because they are more widespread. While there is evidence from both paleontological and extant assemblages that this is indeed so (McKinney 1998), there is also an increasing number of examples of marine species which have recently become extinct or appear to have a high probability of doing so in the near future (Casey and Myers 1998; Carlton *et al.* 1999; Roberts and Hawkins 1999; Hawkins *et al.* 2000). The latter finding is perhaps unsurprising given the intensity of exploitation of many marine species, as well as the indirect consequences thereof (e.g. habitat destruction through fishing activities).

In comparison with both marine and terrestrial species, freshwater species doubtless have ranges that on average are much smaller; the extent of freshwater ecosystems tends to be very restricted (*c.*2.5 × 10^6 km^2, <2 per cent of the Earth's surface; Heywood 1995). This explains, in large part, why extinction rates in freshwater systems tend to be yet higher than those of terrestrial systems; the vulnerability of freshwater environments to point disturbances (e.g. pollution events) and invasive species doubtless also contributes. Ricciardi and Rasmussen (1999) calculate that recent and future extinction rates are higher for freshwater fauna in North America than for terrestrial and marine fauna, with the projected mean future rate for freshwater fauna being five times greater than that for terrestrial fauna and three times the rate for coastal marine mammals. The freshwater rates fall within the range of estimates for tropical rain forest communities.

3.5.3 Biogeographic region

Within the terrestrial realm there is some limited evidence for systematic differences between biogeographic regions and continents in the geographic range sizes of their constituent species (Rapoport 1982; Anderson and Marcus 1992; Smith *et al.* 1994; Allsopp 1999). Thus, Rapoport (1982) reports a positive relationship between the sizes of continents and the mean range sizes of the bird species that occupy them, with the latter measure equating to approximately one-quarter of the former. Anderson and Marcus (1992) report greater geometric mean range sizes in North America than Australia for species of Anura, Reptilia (lizards, snakes, and turtles), Aves, and Mammalia (all species, and separately for bats but not rodents) (Fig. 3.12). They also found a positive relationship between the geometric mean range sizes in Australia and North America of species in different orders of birds, with the most marked deviations from equality lying in the direction of larger ranges in North America. For mammals, Smith *et al.* (1994) find that mean range sizes are smaller for Australia than for North America, and smaller for North America than for the Palaearctic, following the sequence of continental size (see also Letcher and Harvey 1994). However, median range sizes for North America are smaller than those for Australia, which arguably follows from the high topographic relief of the former region.

I am unaware of equivalent published analyses for marine systems, although it seems likely that maximum and mean range sizes of species in individual taxonomic groups will tend to track variation in the sizes of marine provinces and ocean basins.

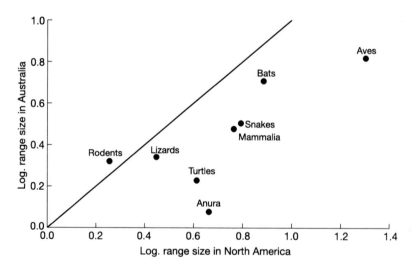

Fig. 3.12 Relationship between the log. geometric mean range size ($10^5 \, km^2$) in North America and Australia of species in different groups of vertebrates. The solid line is the line of equality. From data in Anderson and Marcus (1992).

3.5.4 Latitude

In a stimulating and highly influential paper, Stevens (1989) highlighted the existence of a decline in the geographic extents of the distributions of species from high to low latitudes, christening this as 'Rapoport's rule'; Eduardo H. Rapoport (1982), it was argued, had originally drawn attention to such a trend (although it was actually noted considerably earlier; Lutz 1921). Despite the large number of published studies (Table 3.5; see also Battisti and Contoli 1999), the generality of this pattern has subsequently been much debated, and is clearly much narrower than originally thought (Rohde 1996; Gaston *et al.* 1998b). Indeed, although it is plainly exhibited in some cases, many of which address analytical complexities, a latitudinal increase in range size is only exhibited in a small to moderate proportion and over limited latitudinal spans (see Gaston *et al.* 1998b for a review); this is true, regardless of the measure of range size employed and whether marine or terrestrial data are considered (most discussion of latitudinal patterns in the distributional extents of species has concerned latitudinal extents, but there seems no reason not to generalize the discussion more widely).

There are two basic methods for analysing latitudinal variation in geographic range sizes (Gaston *et al.* 1998b). Stevens' method compares, for different latitudes, the average range sizes of all species residing at each (Stevens 1989). The mid-point method compares, for different latitudes, the range sizes of species whose distributions are centred at each (Rohde *et al.* 1993). The former approach has the disadvantage that a single species can contribute to the mean range size at more than one latitude, and as a result the means for each latitude are not statistically independent. The mid-point method overcomes this problem, because the range of each species is only considered once (at its latitudinal midpoint). Both methods suffer from problems of phylogenetic non-independence of data points, although this is more obvious in the case of the mid-point method, and has sometimes been accounted for in that context. Regardless of their limitations, the two methods provide different information, and neither is *a priori* the more appropriate for most of the analyses that have been conducted to date.

In documenting examples of Rapoport's rule, Stevens (1989) explicitly did not consider data from low latitudes, on the grounds that geographic ranges in these areas were likely to be disproportionately under-estimated as a consequence of sampling biases. Nonetheless, some of the plots he presented hinted at an upturn in range extents towards such latitudes. Subsequent analyses have extended to low latitudes for groups for which the data are likely to be reasonably reliable, and have found this upturn to be genuine, and that in the tropics ranges may on average be larger than in adjacent temperate regions (see Gaston *et al.* 1998b).

For several groups of organisms it has been argued independently, and albeit sometimes in a rather anecdotal fashion, that the geographic ranges of tropical species may be quite large (see Richards 1973; Terborgh 1973; Gaston 1991b; Pitman *et al.* 1999). For terrestrial systems, this is particularly true of tropical Africa, is to a degree also true of the New World, but probably extends less forcefully to the

Table 3.5 Published studies of interspecific relationships between geographic range size and latitude

Taxon	Region	Source
Plants		
trees	North America	Stevens (1989, 1992)
Pinus	North America	Stevens and Enquist (1997)
temperate rainforest trees	South America	Arroyo *et al.* (1996)
Acacia	Australia	Edwards and Westoby (1996)
eucalypts	Australia	Edwards and Westoby (1996), Hughes *et al.* (1996)
columnar cacti	Mexico and Argentina	Mourelle and Ezcurra (1997)
seaweeds	Eastern Atlantic, eastern Pacific	Santelices and Marquet (1998)
Invertebrates		
marine molluscs	North America	Stevens (1989, 1992)
marine molluscs	eastern Pacific	Roy *et al.* (1994)
freshwater crayfish	North America	France (1992)
freshwater amphipods	North America	France (1992)
Brachinus beetles	North America	Juliano (1983)
Anisodactylus beetles	North America and Eurasia	Noonan (1999)
Fish		
freshwater and coastal fishes	North America	Stevens (1989, 1992)
marine teleost fishes	Various	Rohde *et al.* (1993)
freshwater fishes	North America, Northern Europe, Australia	Rohde *et al.* (1993)
marine fishes	East Atlantic	Macpherson and Duarte (1994)
Cyprinella minnows	North America	Taylor and Gotelli (1994)
marine teleost fishes	Indo-Pacific, Atlantic	Rohde and Heap (1996)
marine fishes	Pacific Ocean	Stevens (1996)
Amphibians and reptiles		
amphibians and reptiles	North America	Stevens (1989, 1992)
turtles	global	Hecnar (1999)
Birds		
birds	Soviet Union	Stevens (1989)

(continued)

Table 3.5 *(continued)*

Taxon	Region	Source
non-migratory birds	Soviet Union	Stevens (1989)
birds	North America	Brown (1995, p.113)
birds	New World	Blackburn and Gaston (1996a)
wildfowl	global	Gaston and Blackburn (1996b)
Phylloscopus leaf warblers	Old World	Price *et al.* (1997)
parrots	New World	Koleff and Gaston (2001)
woodpeckers	New World	Koleff and Gaston (2001)
woodpeckers	global	Blackburn *et al.* (1998a)
andean passerines	South America	Ruggiero and Lawton (1998)
Mammals		
mammals	North America	Stevens (1989, 1992)
mammals	North America	Pagel *et al.* (1991)
mammals	Palaearctic	Letcher and Harvey (1994)
mammals	South America	Ruggiero (1994)
mammals	Australia	Smith *et al.* (1994)
mammals	Australia	Johnson (1998b)
bats	New World	Lyons and Willig (1997)
marsupials	New World	Lyons and Willig (1997)
primates	Africa	Cowlishaw and Hacker (1997)
primates	Africa	Eeley and Foley (1999), Eeley and Lawes (1999)
primates	South America	Eeley and Lawes (1999)

Indotropics where the land mass is highly fragmented and where local species in many groups are numerous. Thus, writing of the flora, Richards (1973) observed that 'In the forests of tropical Africa… a very large proportion of the species have wide, though sometimes discontinuous, ranges. Species of endemic or localized distribution are relatively few. This is in sharp contrast to the situation in South and central America and in Malaya, Borneo, New Guinea and other parts of IndoMalaysia where local species are very numerous.'

A common experience amongst tropical biologists is the dramatic expansion of the known geographic ranges of species as sampling improves. What initially may appear to be quite narrow endemics later become understood to have much larger ranges, albeit they may occur at low densities or be patchily distributed. This is not to deny, of course, that the tropics are also areas in which large numbers of species with restricted geographic ranges occur. Thus, Stattersfield *et al.* (1998) observe that the tropics are by far the most important zone for Endemic Bird Areas (an area which encompasses the overlapping breeding ranges of restricted-range bird species—those with range of 50 000 km² or less—such that the complete ranges of two or more such species are entirely included within the boundary of the EBA). On balance, however, the frequency of such species is not sufficient to generate a simple decline in range size from high to low latitudes.

Despite the limited evidence for a Rapoport effect the pattern has received much attention, and rapid entry into popular texts (e.g. Wilson 1992). One reason why this may have been the case is that the mechanism which Stevens (1989) proposed as its cause is intuitively attractive, and has deep historical roots (e.g. Hutchins 1947; Dobzhansky 1950; Janzen 1967; Brattstrom 1968). The climatic variability hypothesis postulates that a latitudinal gradient in range size occurs because towards higher latitudes temporal climatic variability is higher, individual organisms need to be able to withstand this greater variability, and thus the species to which they belong can become more widespread.

The failure of a growing number of studies to document a Rapoport effect across the full span of latitudes might be seen as itself providing a basis for dismissing the climatic variability hypothesis (as might the existence of models which predict the converse latitudinal pattern in range size; Colwell and Hurtt 1994). A climatic variability hypothesis can, however, explain the observed latitudinal patterns in geographic range sizes if Stevens' original formulation is modified somewhat (Gaston and Chown 1999b). Stevens (1989) only considered the effects of variance in climatic conditions on range sizes. However, the mean is also important (Gaston and Chown 1999b). Mean environmental temperatures increase from the Poles until about 25° north or south, at which point they level out (Fig. 3.13; Terborgh 1973). Thus, environmental conditions are reasonably constant across the tropics. Treeline and snowline, for instance, fall at an altitude that is approximately stable throughout this region (Körner 1998). Any species that can endure conditions at these critical points can in terms of broad patterns of climate, potentially, therefore, occupy all latitudes from 25°N to 25°S. The range size of such species is then only limited by the amount of land between these latitudes. In fact, Rosenzweig (1992, 1995) has shown that the land area of the tropics is vastly greater than that of any other biome. Thus, tropical species can have large ranges because the area of the tropics is great, whereas high latitude species can have broad ranges because of their ability to withstand a broad range of conditions (Gaston and Chown 1999b).

If climatic variation is important in determining latitudinal variation in geographic range sizes, then one would expect physiological tolerances to change with latitude accordingly. Gaston *et al.* (1998b) collated some evidence that this was indeed so, and Spicer and Gaston (1999) provide a general review (see also Grace 1987; Woodward 1987; Wiencke *et al.* 1994; Santelices and Marquet 1998; Addo-Bediako *et al.* 2000).

The temporal stability of environmental conditions may, in general, play an important role in influencing the range sizes of species. Thus, for example, running-water (lotic) species of aquatic Coleoptera tend to have smaller ranges than those

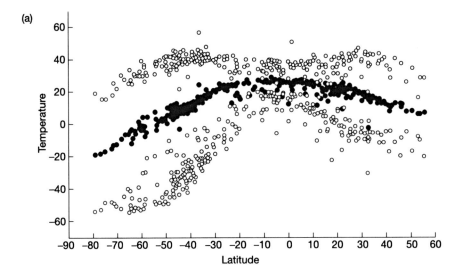

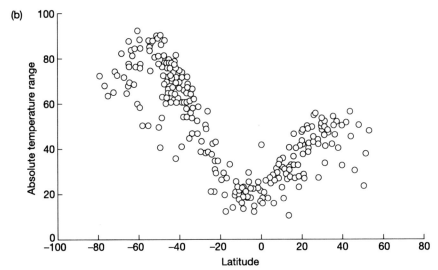

Fig. 3.13 Latitudinal variation in (a) absolute maximum and minimum temperatures (open circles) and annual mean daily temperature (closed circles) (°C), and (b) the range of absolute temperature variation (°C), for mainland locations in the New World (latitude in degrees; northern latitudes are negative). Note that locations vary in altitude, the numbers of years of record and the period that those years embrace. From Gaston and Chown (1999b).

occurring in standing-water (lentic), which has been argued to result from the dynamics of the two habitat types; stagnant water bodies are more likely to completely disappear, requiring frequent migration of resident populations (Ribera and Vogler 2000).

Table 3.6 Mechanisms proposed to explain latitudinal gradients in geographic range size

Climatic variability
Biogeographic boundaries
Differential extinction
Land area
Competition

At least four mechanisms other than that of climatic variability have been proposed to explain Rapoport's rule (Table 3.6; Gaston *et al.* 1998b), based on patterns in biogeographical boundaries, differential extinction, land area, and competition. The first of these proposes that range sizes may be determined by the extent of biogeographic provinces if species can expand their distributions more easily within than across provincial boundaries (Roy *et al.* 1994). This would explain the coincidence between the inflection point in range sizes observed at about 15 to 20°N for some groups in the New World and the high level of floral and faunal turnover at this latitude (Leith and Werger 1989; WCMC 1992; Gauld and Gaston 1995; Blackburn and Gaston 1996a; Ortega and Arita 1998). However, it does raise the question of what determines biogeographic boundaries, and the spectre of circularity if biogeographic regions are defined on the basis of faunal turnover. Moreover, if it could be shown that biogeographical boundaries are areas where changes in environmental conditions limit species distributions, this explanation could largely be subsumed into the modified climatic variability hypothesis, as described above.

Of the other hypotheses for Rapoport's rule, that based on differential extinction (Brown 1995) has the potential to explain why variation in range sizes should differ north and south of the equator. The hypothesis proposes that species at higher latitudes may have larger range sizes because those species in these areas which had narrow tolerances, and hence restricted occurrence, underwent differential extinction due to glaciation and climate change (Brown 1995). A link between glaciation and Rapoport's rule has also been made by Price *et al.* (1997), who examined the geographic range sizes of Palaearctic taxa in the genus *Phylloscopus* (leaf warblers). They suggested that the pattern arises from the different colonization abilities of species invading areas following the retreat of glaciers. Glaciation hypotheses can explain the hemispheric asymmetry in range size variation, because the effects of glaciation were far more severe in the northern than in the southern hemisphere (Markgraf *et al.* 1995). However, it is difficult to see how this explanation can account for the large average ranges of tropical species in many taxa (e.g. Rohde *et al.* 1993; Roy *et al.* 1994; Ruggiero 1994; Blackburn and Gaston 1996a; Gaston and Blackburn 1996b; Lyons and Willig 1997). The competition hypothesis, which suggests that species at higher latitudes may have larger range sizes because of lower levels of competition resulting from the lower species richness (Pianka 1989; Stevens 1996; Stevens and Enquist 1997), falls at the same hurdle. Finally, hypotheses based on variation in land area can be ruled out, as analyses controlling for variation in land area still reveal latitudinal variation in range sizes (Rapoport 1982; Pagel *et al.* 1991; Blackburn and Gaston 1996a). In sum, only the modified climatic variability hypothesis presently stands as an adequate explanation for latitudinal patterns in range size variation.

Increases in geographic range size with latitude have been argued to be associated with higher species richness in tropical regions (Stevens 1989). The lack of any simple directional trend in range size makes such a connection highly unlikely (Roy *et al.* 1994; Gaston and Blackburn 2000; see also Kerr 1999), although clearly there are interactions between patterns of speciation, extinction and range size (Section 3.4).

3.5.5 Longitude

Perhaps in the expectation of finding little pattern, there has been a limited amount of work on variation in geographic range size with longitude, although studies of this kind may help the development of an understanding of why ranges differ in size, particularly because longitudinal gradients do not tend consistently to be associated with systematic variation in environmental parameters as do latitudinal gradients. Certainly there are some longitudinal patterns in geographic range size, although these tend to be less marked and less consistent than latitudinal patterns (some studies fail to find any longitudinal patterns; e.g. Eeley and Foley 1999). Thus, range sizes of mammal species tend to peak in the centre of Australia and decline both to east and west (Smith *et al.* 1994), and to decline from east to west in North America (Pagel *et al.* 1991).

3.5.6 Trophic group

There is a widely held impression that species at higher trophic levels tend, on average, to have larger geographic ranges. In fact, in contrast to the good evidence that predators are rarer than non-predators in terms of numbers of individuals (Spencer 2000), there is little evidence to sustain such a belief. Certainly there are numbers of predators with very small geographic ranges. Farlow (1993) provides a fascinating discussion of the likely sizes of the geographic ranges of tyrannosaurs and other large theropods, which seem on present evidence to have been remarkably small, although the incompleteness of the fossil record and undue taxonomic splitting may give a false impression of just how widely distributed these 'titanic predators' were.

Although there may be no simple patterns in the geographic range sizes of species in different trophic groups, range sizes undoubtedly have implications for interspecific interactions (as well as potentially being determined by those interactions; Section 2.2.4–2.2.6). Thus, the geographic range sizes of host species tend to be positively related to the number of species of consumers that exploit them. Indeed, such a pattern has been documented for fungi and insects on/in host plants (e.g. Lawton and Schröder 1977; Cornell and Washburn 1979; Lawton and Price 1979; Claridge and Wilson 1982; Newton and Haigh 1998; Kelly and Southwood 1999)[6], and

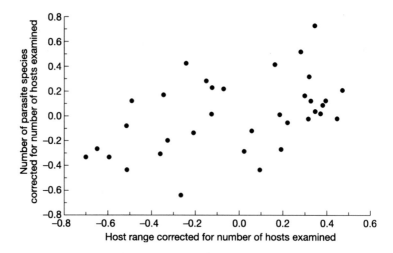

Fig. 3.14 Relationship between the numbers of parasite species and the geographic range of the host species for 35 species of waterfowl hosts. The variables are the residual values derived from regressions of $\log_{10}$ [number of parasite species per host] and $\log_{10}$ [host geographic range] on $\log_{10}$ [number of host individuals examined], to control for problems related to host sampling. From Gregory (1990).

parasites on/in vertebrate hosts (e.g. Price *et al.* 1988; Gregory 1990; Bell and Burt 1991; Poulin 1998). Concerns have been expressed that some examples of these patterns are simply a function of differential sampling effort being directed toward more widespread species, which will also be locally more abundant (Section 3.5.9; Gregory 1990). However, the general patterns seem to be maintained when these effects have been controlled for (Fig. 3.14), and can readily be interpreted in terms of island biogeography, with potential hosts that have larger geographic ranges providing larger targets for colonization (Strong *et al.* 1984). This effect is likely to be enhanced if species with larger ranges tend also to persist for longer, and thus have longer to accumulate colonists.

3.5.7 Body size

A common assertion in the ecological literature is that small-bodied species tend to have smaller geographic range sizes than do large-bodied. In the absence of any qualification it is difficult to judge such a statement. At its broadest, across all organisms, it seems unlikely that there is any simple relationship between geographic range size and body size. Some of the very smallest species (e.g. many microorganisms) and some of the very largest (e.g. some cetaceans) can be very widely distributed. Thus, Fenchel, Finlay and others have argued that the geographic range sizes of protists are commonly extremely large (Fenchel 1993; Finlay and Clarke 1999; Finlay and Fenchel 1999; Finlay *et al.* 1999; Wilkinson 2001). These authors have suggested that the reason for such extensive distributions is that their abundances are exceedingly high, they are probably capable of population growth in a broad range of conditions, and the termination of population growth is accompanied by cyst production, and as a result the global species richness of protists is actually quite low (an issue of particular relevance to understanding the shape of the species–body size distribution for animals; May 1978; Fenchel 1993; Finlay *et al.* 1996, 1999; Finlay and Clarke 1999; Gaston and Blackburn 2000).

Within much narrower taxonomic bounds, such as those of land birds or mammals within a biogeographic region or continent, interspecific relationships between body size and geographic range size appear to exhibit an approximately triangular form (Table 3.7; Brown and Maurer 1987, 1989; Gaston 1994a; Taylor and Gotelli 1994;

Table 3.7 Summary of the findings of studies of the interspecific relationships between geographic range size and body size; '+' positive, '−' negative, and 'NR' no statistically significant relationship; from Gaston and Blackburn (1996c), with additions

Taxon	Region	Relationship	Source
Plants			
palms	Amazon	+	Ruokolainen and Vormisto (2000)
Invertebrates			
mantis shrimps	Global	+	Reaka (1980)
Brachinus beetles	North America	NR	Juliano (1983)
dung beetles	Afrotropics	+	Cambefort (1994)
Fish			
freshwater fishes	West Africa	+	Hugueny (1990)
Cyprinella minnows	North America	+	Taylor and Gotelli (1994)
coral reef fishes	Global	Restricted range species smaller-bodied than non-restricted range species	Hawkins *et al.* (2000)
marine fishes	Atlantic and Mediterranean	+	Rapoport (1994)
suckers	North America	+	Pyron (1999)
sunfishes	North America	NR	Pyron (1999)
Amphibians and reptiles			
frogs	Australia	+	B. R. Murray *et al.* (1998)
turtles	Global	+	Hecnar (1999)

(continued)

Table 3.7 *(continued)*

Taxon	Region	Relationship	Source
Birds			
birds	North America	+	Brown and Maurer (1987)
birds	Australia	+	Maurer *et al.* (1991)
introduced birds	Australia	–	Duncan *et al.* (2001)
Mammals			
cetaceans	Global	+	Rapoport (1994)
Peromyscus mice	North America	+, – and NR	Glazier (1980)
forest mammals	Neotropics	+	Arita *et al.* (1990)
primates	Amazon	+	Ayres and Clutton-Brock (1992)
primates	Global	NR	Gaston (1994a)
primates	Africa	?	Eeley and Lawes (1999)
mammals	North America	+	Van Valen (1973b), Brown (1981), Brown and Maurer (1989), Brown and Nicoletto (1991), Pagel *et al.* (1991)
mammals	Australia	NR	Johnson (1998b)
mammals	Australia	NR (but near-triangular relation-ships for both contemporary and pre-European faunas)	Murray and Dickman (2000)

Blackburn and Gaston 1996a). Here, species of all body sizes may have large geo-graphic range sizes (the upper limit normally being imposed by the size of the study area) but the minimum range size exhibited by species tends to increase with body size. For example, Arita *et al.* (1990) document a broadly positive relationship between geographic range size and body mass for 100 species of Neotropical forest mammals, and Brown (1981) found such a relationship for North American terres-trial mammal species. Figure 3.15a shows the relationship between total geographic range size and body mass in those New World birds not currently threatened with extinction. This can be considered to be triangular, with the upper and left hand edges relatively clearly defined, although statistical testing of such space-filling pat-terns has proven difficult (Blackburn *et al.* 1992; Thomson *et al.* 1996; Scharf *et al.* 1998; Gotelli 2001). How well defined the lower boundary (hypotenuse) to this and

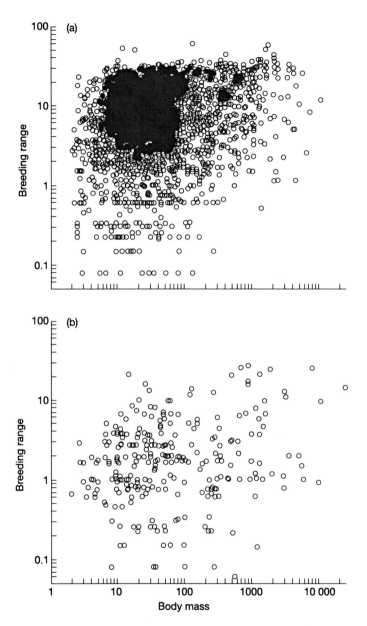

Fig. 3.15 Relationship between breeding geographic range size and body mass (g) for New World endemic landbird species (a) not listed as under threat of extinction and (b) listed as under threat of extinction. From Blackburn and Gaston (1996a).

other triangular range size–body size relationships is considered to be is, to some extent, in the eye of the beholder. Indeed, the relationship in Fig. 3.15a has variously been described as a blob, an arrowhead, a circle, a polygon, or a square. (Note that the bottom left-hand corner of the triangle appears to be cut off, probably because of an artefact of the methodology used to plot this relationship; minimum geographic range is set by the grid-based method used to quantify the range sizes of New World birds; Blackburn and Gaston 1996a.) Certainly, within many groups there are, or have in the past been, very large-bodied species that do not seem to have had particularly large, natural geographic range sizes. For example, the largest known eagle, Haast's eagle *Harpagornis moorei* (9–13 kg), now extinct, was apparently restricted to New Zealand. Indeed, many examples of gigantism derive from endemics to these islands (Daugherty *et al.* 1993).

Although difficult to demonstrate empirically at present, I suspect that, in general, more obviously triangular relationships, as opposed to a complete absence of pattern, are more readily recovered from comprehensive analyses of data for vertebrate than for invertebrate groups. More studies of the latter would, however, be necessary to substantiate this point.

Whilst they may typify comprehensive analyses (*sensu* Section 1.3), triangular patterns of relationship between range size and body size are not typical of partial analyses (Gaston and Blackburn 1996a). The latter fail to document any reasonably consistent pattern, with positive, negative, and the complete absence of relationship having been reported. Even for entire geographic ranges, it seems reasonable to expect relatively poor relationships between range size and body size within taxonomic groups, given that many pairs of sister species of similar body size but very different range sizes can readily be identified—this fits with different genetic constraints on range size and body size (Section 3.5.1). Moreover, the existence of a strong pattern would tend to be opposed by any tendency for smaller species to have higher abundances (see Gaston and Blackburn 2000 for a review) and for higher abundances to be associated with larger range sizes (Section 3.5.9). Nonetheless, five mechanisms have been suggested to explain a positive interspecific range size–body size relationship within taxonomic assemblages at the scales of comprehensive analyses:

1. *Phylogenetic non-independence* (Gaston and Blackburn 1996b)—an artefactual relationship between range size and body size could result from the shared common ancestries of species in an assemblage. Because of their phylogenetic relatedness, species do not constitute independent data points for analysis, inflating the degrees of freedom available for testing statistical significance (Harvey and Pagel 1991; Harvey 1996). If sufficient, this inflation may falsely imply that relationships exist which in reality do not, although it is not obvious that such an effect would predict a triangular range size–body size relationship.

   In the main, analyses of range size–body size relationships have taken no account of possible phylogenetic effects. However, there are a growing number of exceptions. Thus, Glazier (1980) finds that wide-ranging species of *Peromyscus* mice tend, on average, to have smaller body sizes, but that the trend differs between species groups, such that for *Haplomylomys*, *Truei*, and *Mexicanus* it is negative, in *Maniculatus* it is positive, and for *Leucopus*, *Boylei*, and *Melanophrys* no trend is evident. Taylor and Gotelli (1994) find, for a comprehensive scale analysis, that when the effects of phylogenetic relatedness are removed a positive relationship between range size and body size for *Cyprinella* minnows remains. Gaston and Blackburn (1996b) found that there was no significant relationship between geographic range size and body size for wildfowl at a global scale, whether or not the phylogenetic relatedness of species was controlled for. Similarly, the weak positive relationship across species of New World birds disappeared when the effects of phylogeny were controlled for (Blackburn and Gaston 1996a); there was also no significant tendency for positive range size–body mass relationships to outnumber negative ones within New World bird taxa, so that the failure to recover a relationship within taxa is not a result of its non-linearity across species. Pyron (1999) documents positive relationships between body size and geographic range size for sucker and sunfish species, after controlling for phylogenetic relatedness. In short, controlling for phylogeny seems to have no consistent effect on whether a range size–body size relationship is observed, and if so whether this is the same relationship as is seen without so doing.

2. *Minimum viable population size* (Brown and Maurer 1987; Brown 1995; see also Marquet and Taper 1998)—on average, larger-bodied species may have, or may require, larger-sized home ranges than smaller-bodied ones (McNab 1963; Armstrong 1965; Schoener 1968; Clutton-Brock and Harvey 1977; Linstedt *et al.* 1986; Swihart *et al.* 1988; Peery 2000). This being so, they may also require larger total geographic range sizes in order to maintain range-wide minimum viable population sizes. This would tend to result in a positive interspecific range size–body size relationship, and indeed a triangular one, because there is no necessary upper constraint on the range size of small-bodied species.

   This has been the generally favoured mechanism for positive triangular body size–range size relationships, principally because it seems logically sound and plausible. Certainly, many species that are at present regarded as at threat of extinction in the near future have geographic ranges that are smaller than otherwise expected for their body size (Fig. 3.15).

3. *Realized/potential geographic range size ratios* (Gaston and Blackburn 1996a, c)—minimum range size may increase with body size because, on average, larger-bodied species disperse more rapidly and successfully (i.e. manage to establish) than small-bodied, or are evolutionarily older and have therefore had longer periods in which to disperse, establish, and attain larger geographic range sizes. Larger-bodied species may therefore occupy (realize) a greater proportion of their potential range sizes (the geographic range size which could be occupied were all barriers to dispersal to be overcome; Gaston 1994a). This may generate a positive range size–body size relationship, and may generate a triangular one if

some small-bodied species occupy a high proportion of their potential range sizes, but the average small-bodied species occupies a smaller proportion than the average large-bodied species.

There is no evidence to support this hypothesis, although Hausdorf (2000) argues that a negative correlation between the geographic range size of families and the median shell diameter of their constituent genera for Limacoidea snails indicates the greater role of passive dispersal (which decreases with body size in this group) than active dispersal (which increases with body size) in generating variation in distributional extents. For groups which spread by passive dispersal this tends to be more typical of the smaller-bodied species.

4. *Homeostatic and environmental variabilities* (Gaston 1990; Root 1991)—species of large body size are able to maintain homeostasis over a wider array of conditions than are small-bodied species (Boyce 1979; Cawthorne and Marchant 1980; Lindstedt and Boyce 1985; Zeveloff and Boyce 1988; Pimm 1991). If larger geographic ranges encompass greater environmental variability, then it may only be the larger species which are able to maintain them. This would seem to predict a positive range size–body size relationship, although one can envisage circumstances under which this would become triangular, as with mechanism (3).

Root (1991) drew some support for this hypothesis from the observation that in a select group of wintering North American passerines larger species were found to extend their ranges further north than smaller ones, and that northern cardinals *Cardinalis cardinalis* were close to their maximal ability to maintain thermal homeostasis at their northern range limit.

5. *Latitudinal gradients* (Pagel *et al.* 1991)—if the average body sizes and geographic range sizes of species in assemblages both tend to decline towards lower latitudes (Bergmann's and Rapoport's rules, respectively; Bergmann 1847, cited in James 1970; Stevens 1989; Blackburn and Gaston 1996a, b), this could result, perhaps incidentally, in an observed increase in minimum geographic range size with increasing body size. It could also thus result in an overall positive triangular range size–body size relationship.

The limited evidence for a systematic decline in range size with latitude (Section 3.5.4) suggests this mechanism is unlikely to provide a general explanation for a positive range size–body size relationship. The occurrence of clines in both variables may nonetheless occur across some latitudinal spans, requiring in these cases an explanation of the clines to explain the relationship.

On present evidence, the popularity of the minimum viable population size mechanism seems justified. However, the five mechanisms are not mutually exclusive and could potentially be reinforcing. Indeed, they need not be independent. For example, interspecific differences in homeostatic variability (mechanism 4) have been postulated to explain latitudinal trends in the sizes of the geographic ranges of species (mechanism 5; Stevens 1989) and potential range sizes (mechanism 3) have been argued to be smaller in the tropics. Likewise, mechanisms 2 and 4 may be linked, if the only large-bodied species able to attain the large geographic range sizes necessary to maintain minimum viable population sizes are those able to maintain homeostasis over a wide array of conditions.

The effects of any of the mechanistic processes need only be relatively weak to generate the typically poor positive interspecific range size–body size relationships. This emphasizes the fact that some large-bodied species may have small ranges, and not necessarily as a consequence of human activities.

3.5.8 Dispersal ability

Whilst measures of dispersal ability tend in general to be rather crude, a number of studies have sought to test for an interspecific relationship between dispersal ability and geographic range size, with better dispersers being predicted to be more widespread. In plants, no simple picture seems to emerge, although some groups of species with apparently good dispersal abilities may be widespread (e.g. Edwards and Westoby 1996; Kelly 1996; for partial analyses see Peat and Fitter 1994; Kelly and Woodward 1996). In terrestrial and freshwater animal groups, evidence for a positive relationship between dispersal ability and geographic range size or regional occupancy has been claimed on a number of occasions (e.g. Juliano 1983; Kavanaugh 1985; Gutiérrez and Menéndez 1997; Dennis *et al.* 2000; Malmqvist 2000). In the main, for practical reasons these studies rely on traits thought to correlate with dispersal ability rather than actual measures of that ability (e.g. whether winged or not, or whether other dispersal structures are well developed, rather than mark-release recapture studies, or studies of patterns of invasion or appearance in areas). These proxies are doubtless far better in some instances than in others (in some they are probably appalling), with their effectiveness depending particularly on a good understanding of the biologies of the taxa of concern. In some groups, one may need to distinguish carefully between physical and behavioural aspects of dispersal ability. Many forest birds, for example, appear quite capable of dispersing long distances, but may not do so because of a reluctance to cross clearings and other habitat discontinuities.

Amongst birds, migratory species tend to have smaller geographic range sizes than non-migratory, even after controlling for latitudinal effects and local abundance (Bensch 1999). This has been argued to result from the constraints on colonization of new breeding areas by migratory species that results from their complex migratory behaviour (Böhning-Gaese *et al.* 1998; Bensch 1999).

In marine systems, differences in the mode of larval development of invertebrate species have been argued to be related to geographic range size, with indirect development and long larval periods (a form of dispersal) thought to produce large ranges. There is substantial evidence in support of this (e.g. Thorson 1950; Hansen 1978; Jablonski 1986c; Emlet 1995), but some contrary findings and much debate (e.g. Levinton 1988; Palumbi 1994; Emlet 1995). Interpretation of these relationships may be complicated by latitudinal gradients in forms of larval development and geographic range size.

Whether in marine, freshwater, or terrestrial systems the relation between range size and dispersal is nearly always difficult to assess, because for any given species dispersal distances are almost invariably strongly right skewed (e.g. Paradis *et al.* 1998). Most individuals disperse only short distances, but a few may disperse over distances that are orders of magnitude greater. Theoretical and empirical evidence shows that the latter may be of fundamental importance in the colonization and establishment of populations in previously unoccupied areas, and in maintaining gene flow between disparate populations (for which very low rates of immigration, of the order of one individual per generation, may be sufficient). Unfortunately, direct observations of rare long-distance dispersers is typically very difficult.

If we accept that there is a causal link, then the core issue in understanding relationships between dispersal ability and geographic range size is which variable is cause and which is effect. A limited dispersal ability could restrict the range of a species, and in that sense could generate rarity; the ranges of species would be dispersal limited (Section 2.3.2). However, the selective advantages of developing enhanced dispersal abilities would presumably be large if unoccupied habitat and otherwise unexploited resources would become available to individuals. Dispersal ability is probably a relatively plastic trait that can readily be reduced and improved under suitable selective pressures (with the possible exception of the consequences of the entire loss of dispersal structures); Cody and Overton (1996) found that reduction in dispersal ability took place over just a few generations in some weedy, short-lived, and wind-dispersed plants of inshore islands in British Columbia. It is therefore likely that the possession of a small range size for other reasons selects for a reduction in dispersal ability, because there is no advantage to be gained by its maintenance and perhaps some cost, and because its maintenance prevents allocation of energy into other paths that would enhance fitness. It is particularly easy to see that loss of dispersal ability in certain situations, such as on small islands and mountain tops, might be strongly selected for if their maintenance demands resources and their use carries risks of being swept into unoccupiable (and perhaps fatal) territory (for examples of costs of maintaining dispersal abilities see Tanaka and Suzuki 1998; Zera and Brink 2000).

The prediction has been made that species that have large native geographic ranges will be more likely to be successful in colonizing a new area on their introduction to it, perhaps in part because of greater dispersal abilities, and therefore likelihood of finding suitable habitat after introduction. A number of studies have found evidence for such an effect for groups of plants, insects, and birds (e.g. Forcella and Wood 1984; Forcella *et al.* 1986; Moulton and Pimm 1986; Crawley 1987; Hanski and Cambefort 1991; Roy *et al.* 1991; Blackburn and Duncan 2001b; Duncan *et al.* 2001), although Lockwood *et al.* (1999) found no evidence for passeriform species introduced on four oceanic islands (Oahu, Tahiti, St Helena, Bermuda).

3.5.9 Abundance

In stark contrast to the other patterns of variation in geographic range size that have been mentioned, that with abundance appears to be pervasive. For a given taxonomic assemblage, the species with the largest total populations tend to be the more widespread, whilst those with the smallest populations tend to be more restricted (the shape of species–range size distributions for geographic ranges is also very similar to that for species–abundance distributions based on large regional or global population sizes; Gaston and Blackburn 2000). Thus, for example, amongst the wildfowl (Anseriformes), the mallard *Anas platyrhynchos* is globally the most abundant and the most widespread species, whilst the Brazilian merganser *Mergus octosetaceus* is amongst the least abundant and the most restricted (Fig. 3.16). This is not, however, a matter of the kind of straightforward incremental increase in total population size in direct proportion to range size which one might expect. Rather, population size increases faster than range size, such that there is a positive interspecific relationship between local density and range size. Indeed, a tendency for the locally more abundant species in an assemblage also to be the more widely distributed, and the locally rare to have restricted distributions, appears to be amongst the most general of macroecological patterns (Fig. 3.17; Hanski 1982; Brown 1984; Hanski *et al.* 1993; Gaston 1994a, 1996d; Gaston *et al.* 1997a); more widespread species tend also to have greater numbers of local populations (Brown and Briggs 1991).

An interspecific relationship between local abundance and occupancy has been documented at a spectrum of scales, across a diversity of habitats, and for taxa as varied as plants, spiders, bryozoans, grasshoppers, scale insects, hoverflies, bumblebees, macro-moths, butterflies, beetles, bracken-feeding insects, frogs, birds, and mammals[7]. However, only a remarkably small proportion of these analyses concern the entire, or the majority of, the geographic ranges of species (see Table 3.8).

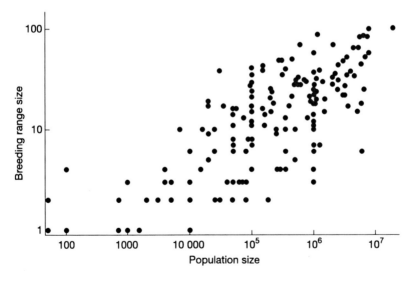

Fig. 3.16 Relationship between global population size (upper estimate) and breeding geographic range size (number of 611 000 km² grid squares occupied) for wildfowl (Anseriformes) species. Based on data reported in Webb *et al.* (2001).

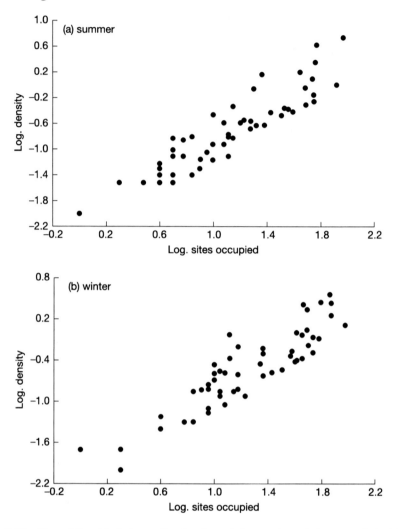

Fig. 3.17 The relationship between the density and the number of sites occupied of bird species in the lowlands of south-east Queensland in (a) summer and (b) winter. Species are only included if present at ≥ five sites across summer and winter. From data in Catterall *et al.* (1998).

In a review of published abundance–occupancy relationships, around 80 per cent were significantly positive (Gaston 1996d). In general, however, the variance explained by the relationship is not very high, of the order of 20 to 40 per cent, although it may reach 80 per cent or more (Gaston 1996d). Indeed, some of the most abundant species are not the most widespread, and some of the most widespread are not the most abundant; the red-billed quelea *Quelea quelea* is widely quoted as being the most abundant, extant, wild bird species in the world (e.g. Wood 1982;

Table 3.8 Summary of the findings of studies of the interspecific relationship between local abundance and geographic range size (i.e. only those analyses of abundance–occupancy relationships concerned with geographic ranges); '+' positive, '–' negative, and 'NR' no statistically significant relationship

Taxon	Region	Relationship	Source
cheilostome bryozoans	Britain	+	Watts *et al.* (1998)
lycaenid butterflies	North America	NR (but local abundance only determined at sites in Colorado)	Hughes (2000)
suckers	North America	+	Pyron (1999)
frogs	Australia	+	B. R. Murray *et al.* (1998)
birds	North America	+	Brown and Maurer (1997)
wildfowl	Global	+	Gaston and Blackburn (1996b)
mammals	Australia	+	Johnson (1998b)
primates	Africa	? (triangular relationship)	Eeley and Lawes (1999)

Campbell and Lack 1985), but is restricted in distribution to the African continent, and conversely none of the most widely distributed bird species mentioned earlier (Section 3.1) would rank amongst the globally most abundant. The strength of abundance–occupancy relationships appears to decline with increasing taxonomic diversity and toward broader spatial scales (i.e. as regional occupancy converges on total geographic range size), which tend to go hand in hand because assemblages are more speciose over larger areas. Such a decline is, nonetheless, expected as a result of an increase in spatial scale alone (He and Gaston 2000).

A number of mechanisms have been proposed to explain interspecific abundance–occupancy relationships in general, although they are not all mutually exclusive (Table 3.9; Gaston *et al.* 1997a). These have been discussed at considerable length elsewhere (Hanski *et al.* 1993; Gaston *et al.* 1997a, 2000; Gaston and Blackburn 2000). Below, the most salient points with regard to abundance–range size relationships are extracted for each mechanism in turn, with the exception of an argument centred on spatial patterns of aggregation of individuals, discussion of which will be deferred until Section 4.3.

1. *Sampling artefact* (Brown 1984; Wright 1991)—a positive interspecific abundance–range size relationship will result simply from the systematic under-estimation of the range sizes of those species with lower local abundances. These stand a greater chance of being missed from samples because they are locally rare. There have been few explicit attempts to control for such an effect in analyses, and to

Table 3.9 Mechanisms proposed to explain positive interspecific relationships between local abundance and geographic range size (although variation in range position might provide a possible explanation for abundance–occupancy relationships more generally, it is not relevant to entire geographic ranges)

Sampling artefact
Phylogenetic non-independence
Niche breadth
Niche position
Density dependent habitat selection
Metapopulation dynamics
Vital rates
Latitude

---

determine its magnitude. It is generally accepted that for many of the data sets that have been examined the spatial distributions of species are sufficiently well known that this bias will not be sufficient to generate the observed pattern. This is not to say, however, that the effect is unimportant, particularly where sampling effort is poor (Watts *et al.* 1998).

2. *Phylogenetic non-independence* (Gaston *et al.* 1997a)—an abundance–range size relationship could result from the phylogenetic non-independence of species as data points for statistical analysis (as with range size–body size relationships; Section 3.5.7). The few analyses that have controlled for the effects of phylogenetic non-independence have found that the abundance–range size relationship is still recovered (Blackburn *et al.* 1997b; Gaston *et al.* 1997c; Quinn *et al.* 1997b; B. R. Murray *et al.* 1998). It is therefore unlikely that this mechanism contributes significantly to the observed pattern. This would follow from the fact that neither geographic range size nor abundance exhibit marked phylogenetic constraint (Section 3.5.1).

3. *Range position* (Bock and Ricklefs 1983; Bock 1984)—if species closer to the edges of their geographic ranges have lower abundances in, and occupy a smaller proportion of, study areas, then a positive abundance–occupancy relationship could result. However, such a mechanism cannot explain relationships for entire geographic ranges.

4. *Niche breadth* (Brown 1984)—if species with greater niche breadths are able, in consequence, to attain higher local abundances and wider distributions then positive abundance–range size relationships should result. Many published studies bear on this hypothesis, testing for relationships between niche breadth and abundance and/or range size. However, in general, when sampling effort has been controlled for (which it often is not), these fail to find significant correlations between breadth of resource (food or habitat) use and local abundance and/or geographic range size (Gaston *et al.* 1997a; Gregory and Gaston 2000;

Hughes 2000). Positive relationships seem to occur more frequently between breadth of resource use and geographic range size than between breadth of resource use and local abundance; however, analyses of the former are particularly prone to being confounded by differences in sampling effort (with the resource breadth of more widely distributed species being estimated on the basis of more individuals or individuals dispersed over a wider area). Although there are hazards in relying heavily on pair-wise comparisons in isolation, Walck *et al.* (2001) provide a nice example of such difficulties. Comparing two species of Asteraceae, the narrowly endemic *Solidago shortii* and the geographically widespread *S. altissima*, they observe that because it has a larger geographic range the latter occurs in more physiographic regions, climatic provinces, hardiness zones, and vegetation types, over a wider range of altitudes and soil types. However, *S. shortii* has been cultivated out-of-doors well north and west of its present range and thus in other physical and biotic regions, both species grow over a wide range of soil fertilities, and in the green-house *S. shortii* grew on a wide diversity of soil types.

5. *Niche position* (Hanski *et al.* 1993; Gaston *et al.* 1997a)—if species with lower niche positions (*sensu* Shugart and Patten 1972), that is they use those resources more typical of a study area or region, are, as a result, able to become locally more abundant and more widely distributed, then a positive abundance–range size relationship could be generated. There is growing support for this hypothesis, resulting from empirical demonstrations that resource availability is a strong correlate both of local abundance and range size (Järvinen and Väisänen 1979; James *et al.* 1984; Burgman 1989; Pearson 1993; Gaston *et al.* 1997a; Gregory and Gaston 2000).

6. *Habitat selection* (O'Connor 1987)—density-dependent habitat selection (driven through intraspecific competition) can, in principle, give rise to positive interspecific abundance–range size relationships, if the intraspecific abundance–range size relationships generated by this selection are sufficiently similar for different species, and such selection operates on a sufficiently large scale. Both of these conditions seem unlikely to apply, although at a smaller spatial scale species may commonly occupy more habitats when densities are high and less when they are low (Kluyver and Tinbergen 1953; Fretwell and Lucas 1970; for a review see Rosenzweig 1991).

7. *Metapopulation dynamics* (Hanski 1991a, 1991b; Gyllenberg and Hanski 1992; Hanski and Gyllenberg 1993; Hanski *et al.* 1993)—a positive abundance–range size relationship is a prediction of metapopulation structures of the form of the rescue effect hypothesis. This assumes that immigration decreases the probability of a local population going extinct (the rescue effect; Brown and Kodric-Brown 1977), and that the rate of immigration per patch increases as the proportion of patches that are occupied increases. Here, for many parameter values, a positive relationship between local abundance and number of occupied patches can result. Again, however, it seems unlikely that such processes operate on a sufficiently

large scale to generate positive interspecific relationships between abundance and geographic range size; if species do exhibit metapopulation dynamics (for which there seems to more evidence for plants, insects, and small mammals than for birds and large mammals), it seems likely that once ranges become moderate to large in size they will comprise multiple metapopulations.

8. *Vital rates* (Holt *et al.* 1997)—if movement between sites is just sufficient that all potentially occupiable sites are occupied, but is not so great that immigration significantly affects the population supported at a site, then it follows that a species will only persist at sites at which its birth rate, b, exceeds its density-independent death rate, d. In other words, since $b - d = r$, a species' intrinsic rate of increase at a site, a site is occupied when $r > 0$. A species will occupy more sites when its average r is high, and fewer when average r is lower. In addition, population dynamic theory predicts that the abundance attained by a species at a site will be proportional to r, assuming that the effect of density dependence is uniform across all sites. Therefore, both the number of sites occupied and mean abundance at those sites will depend on the species' value of r, and hence on the relationship between birth and death rates across sites. It follows that temporal fluctuations in birth and/or death rates will lead to synchronized variation in both the abundance attained by a species at different sites, and the number of sites occupied. A positive abundance–occupancy relationship will result (Holt *et al.* 1997). There are too few data with which to evaluate this model; however, it is very robust, with even marked variance in birth and death rates failing to destroy the positive abundance–range size relationship (A.M. Brewer, unpublished analyses). This model will be examined in further detail in the context of the abundance structure of ranges (Section 4.5).

9. *Latitude*—a positive abundance–range size relationship could be generated if local abundance and range size both declined towards lower latitudes, irrespective of whether there was any causal link between the two variables. However, whilst the limited data available suggest that density may well decline towards the equator (Gaston and Blackburn 2000), plainly range size exhibits a rather more complex pattern (Section 3.5.4).

These mechanisms have obvious affinities with those that have been discussed with regard to other patterns in geographic range sizes discussed in this chapter. Indeed, there will be some strong empirical and theoretical connections, and the ultimate goal is a holistic mechanistic understanding, some features of which are already emerging (Brown 1995; Gaston and Blackburn 2000; Blackburn and Gaston 2001).

The present state of understanding of the mechanisms underlying observed interspecific abundance–occupancy relationships is highly unsatisfactory, in as much as whilst a number of potential explanations have been established it has proven difficult categorically to dismiss most of them on either theoretical or empirical grounds. One might therefore conclude either that the tests performed thus far have been inadequate, or that indeed more than one and perhaps several mechanisms do contribute to the pattern. Whilst it would not be difficult to improve testing of virtually all of the mechanisms, through examination of a greater diversity of data sets and a more comprehensive set of analyses, it seems likely that several will withstand such scrutiny. This suggests that several potentially mutually reinforcing mechanisms might generate interspecific abundance–occupancy relationships. If so, this would run contrary to the notion that explanations for very general ecological patterns will tend to be reasonably simple and singular (MacArthur and Connell 1966).

Regardless of its determinants, the interspecific abundance–range size relationship is increasingly seen as central to an understanding of other major macroecological patterns. In particular, attention is focusing on its interaction with the species–area relationship (e.g. Hanski and Gyllenberg 1997; Leitner and Rosenzweig 1997; Ney-Nifle and Mangel 1999; Harte *et al.* 2001), and thence its implications for levels of species loss associated with habitat destruction and fragmentation (Ney-Nifle and Mangel 1999).

3.5.10 Genetic variation

Species with small geographic ranges are predicted to have low levels of genetic variation, per population or overall, compared with more widespread species (particularly congeners, with which they are likely to exhibit greater similarity in biological characteristics). Evidence that this is indeed so has been reported in a number of studies (Fig. 3.18; Karron 1987; Ellstrand and Elam 1993; Lavie *et al.* 1993; Baskauf *et al.* 1994; Frankham 1996; Premoli 1997), although counterexamples have also been documented and not all restricted species are genetically impoverished (e.g. Karron 1987). The heterogeneity in the levels of genetic variability exhibited by narrowly distributed species will rest heavily on the diversity of factors that have led to their small ranges, including how recently this state has been attained, and whether the species have passed through genetic bottlenecks (Karron 1987).

It seems to remain an open question whether or not, on average, the scaling of genetic variation with geographic range size takes the form that might be expected on the basis of the interspecific relationship between population size and range size.

There is a close parallel between the lower levels of genetic variation exhibited by narrowly distributed compared with widely distributed species, and the lower levels of such variation that are often exhibited by peripheral compared with more central populations of an individual species (Section 2.4). There are other such parallels between interspecific and intraspecific patterns. Thus, whilst widespread species are typically locally abundant (Section 3.5.9), individual species exhibit abundance–occupancy relationships (Section 4.3). In both cases, one can see how the interspecific and intraspecific patterns could plausibly be related through the process of speciation.

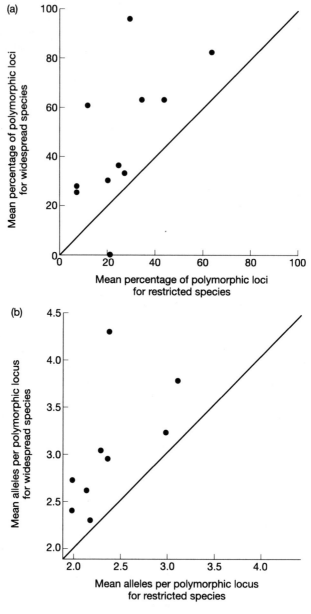

Fig. 3.18 Comparison of (a) the mean percentage of loci exhibiting polymorphism, and (b) the mean number of alleles per polymorphic locus, for restricted and widespread congeneric species in genera of vascular plants. Solid lines are lines of equality; widespread species have significantly higher values than restricted species for both variables. From Karron (1987).

3.6 In conclusion

A conspicuous omission from the discussion of the sizes of geographic ranges provided thus far in this chapter has been any explicit consideration of what determines this area for a given species. Answering this fundamental question requires us to draw on material presented both in this chapter and in its predecessor (Chapter 2). In one sense, of course, the answer is the same as to the question of what determines the limits to the distribution of a species. It can thus be framed in terms of abiotic/biotic factors, population dynamics, and genetics. Indeed, the material discussed in this chapter can be organized under much these same headings.

First, the size of a geographic range can be seen as a reflection of the capacity of a species to overcome spatial and temporal patterns of variation in abiotic and biotic factors. The size is limited by the point at which in each and every direction the levels (mean, variance, etc.) of one or more of these factors, or some combination thereof, prevent the occurrence of populations. The fact that most species have rather small geographic ranges (Section 3.4) suggests that overcoming variation in abiotic and biotic factors is not easy, and the capacity of perhaps the majority of species to do so is quite constrained.

It is a beguilingly easy step from such a picture to the assumption that more widespread species must therefore have greater tolerances of abiotic or biotic factors than those with more restricted distributions. However, whilst, on average, environmental conditions tend to change systematically with latitude and longitude, similar sets of such conditions may be very widely distributed, particularly at the level of the microenvironments that individuals may actually experience. Thus, individuals of even widespread species may only occur under quite a narrow range of environmental conditions by, for example, adjusting the depth or elevation at which they occur with changes in latitude and longitude. Indeed, there is rather little evidence that more widespread species occur in more diverse sets of conditions, but there is growing evidence that they occur under sets of conditions that themselves occur more widely (Section 3.5.9). Indeed, in terms of limiting factors, the best explanation of the fact that most species have small geographic ranges and only a few have large ones seems to be that most sets of environmental conditions are rather narrowly distributed and that only a few themselves occur very widely. Of course, some species are able to exploit much wider variation in environmental conditions than are others, but this does not seem to have a sufficient effect on range size to overwhelm the effect of the availability of appropriate conditions. One might argue that the frequency distributions of range sizes are more reflective of environmental structure than of the environmental capacities of species. Such a pattern of differences in the availability of different sets of conditions seems likely to be predicted by many models of patterns of geographic variation in these conditions.

Second, the size of a geographic range can be seen as a product of population dynamics. It represents the area over which the rate of establishment of local populations is not exceeded by the rate of extinction, with its bounds being defined by the point at which this ceases to be so (see Section 2.3). This area is larger for those species that attain higher local densities (Section 3.5.9) which will, all else being equal, reduce the likelihood of local extinction, increase the cohesion of the range (through the greater number of potential emigrants per unit area), reduce the likelihood of speciation (Section 3.4.1) and thereby increase the likelihood of range size being maintained and also the persistence of the species. Higher local densities can result in larger geographic ranges through very general colonization/extinction dynamics (Warren and Gaston 1997).

Third, and finally, the size of a geographic range can be viewed as a reflection of the genetics of a species. This includes the capacity of individuals to cope with different environmental conditions, the variation in this capacity both within and between populations, and the spatial structuring of this variation. More widespread species tend, on average, to have greater genetic variation both within and between populations, with systematic differences in genotypes in different parts of the geographic range.

4

Abundance structure

The north, south, east, and west boundaries of a species' range tell us very little about what is happening inside....

R. H. MacArthur (1972, p.149)

4.1 Introduction

Position and size are two important descriptors of the structure of the geographic range of a species (Chapters 2 and 3). Nonetheless, they are but crude indices of the distribution of individuals across the land (or sea) scape. Whilst they are all that have been determined about the ranges of many species—and for the majority not even these quantities have as yet been measured (a high proportion of species that are doubtless much more widespread remain known from only a single locality; Andersen *et al.* 1997; Stork 1997)—they entirely ignore any patterns of spatial distribution of individuals that occur within the geographic limits to their occurrence or crude scale occupancy. Knowledge of this more detailed abundance structure of geographic ranges is central to any general theory of the spatial distributions of species. Indeed, this issue has already been alluded to on a number of occasions in this book.

This chapter addresses four principal patterns in the abundance structure of the geographic ranges of species, the form those patterns take, the links between them, and the mechanisms that may be responsible for their existence. The first of these patterns is the intraspecific abundance distribution, the frequency with which a species attains different levels of abundance in space. The second is the intraspecific abundance–range size relationship, the relationship between the mean local abundance that a species achieves and how widespread it is (the intraspecific equivalent of the interspecific relationship addressed previously; Section 3.5.9). The third pattern is that demonstrated by variation in the abundance of a species along environmental gradients, and the fourth is the form of the full pattern of variation in abundance, the abundance surface, that a species exhibits over its geographic range. The first three provide much of the basis for discussion of the fourth.

Curiously, whilst all four patterns constitute fundamental descriptors of the structure of geographic ranges, to date, they have largely been considered in rather strict isolation, and the associated literatures that have developed have primarily been concerned with distinctly different issues. Thus, consideration of intraspecific abundance distributions has centred on identifying the most appropriate statistical models to describe them, study of intraspecific abundance–range size relationships has emphasized their potential implications for resource management (particularly pest control) and conservation, investigation of changes in abundance along environmental gradients has foremost concerned the methodological consequences for ordination analyses, and research into the abundance surface of geographic ranges has focused on the adequacy of its explanation in terms of an optimum response surface model. In this chapter, some of the basic links between these patterns will be emphasized.

Where the ecological mechanisms underpinning the different patterns have been addressed, single models tend to have attracted the vast majority of attention. Again rather curiously, however, these models have been somewhat different in the case of each of the patterns, although the patterns by definition must themselves be closely related. Here, some of these mechanistic relationships will also be highlighted.

As observed earlier, the determinants of the range limits of a species have been argued to be the same as those that determine the levels of abundance within its geographic range. Thus, some of the themes and issues addressed here will be familiar from previous chapters.

4.2 Intraspecific abundance distributions

Perhaps the simplest question one can ask of the abundance structure of the geographic range of a species is what form is taken by the frequency distribution of local abundances, disregarding the spatial relations of the localities. In other words, what is the relative frequency across space with which a species attains different levels of abundance. Investigation of intraspecific abundance distributions has a long history, particularly in connection with problems in applied entomology, although perhaps the majority of literature on the topic is now sadly neglected. Almost universally, these distributions are strongly right or positively skewed. That is, a species is, in relative terms, locally rare in most of the places in which it occurs and locally abundant at relatively few (it is, of course, entirely absent from the vast majority of sites!; e.g. Anscombe 1948; Taylor *et al.* 1978; Perry and Taylor 1988; Mehlman 1994; Brown *et al.* 1995; Gregory *et al.* 1998)[1]. This is true whether, in absolute terms, the mean abundance of the species is high or low, and whether abundance is expressed in terms of local population size or density.

Most studies of spatial variation in the abundances of species concern rather small spatial scales, such as the distribution of individuals of an insect pest between crop plants in a field or between samples taken from adjacent fields. Others have been concerned with animal group formation (e.g. herds, schools, flocks), and the frequency of groups of different size (e.g. Bonabeau *et al.* 1999; Sjöberg *et al.* 2000). Nonetheless, a few studies have considered patterns of variation in population size,

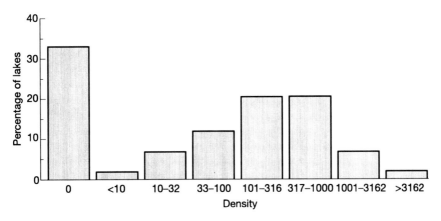

Fig. 4.1 Intraspecific abundance distribution for adults of the zebra mussel *Dreissena polymorpha*, based on lake-wide densities (adults/m^2) in lakes within its European range. From Strayer (1991).

or more usually density over broad areas (Figs 4.1 and 4.2; e.g. Perry and Taylor 1988; Strayer 1991; Brown *et al.* 1995; Winston and Angermeier 1995; Gregory *et al.* 1998; Brewer and Gaston 2002)[2]. Of these, some of the most significant contributions have arisen from the work of the Rothamsted Insect Survey, which has employed networks of light traps and suction traps across Britain to record the abundances of moths and aphids on an essentially continuous basis for several decades (Taylor 1986). These almost invariably reveal that species are relatively scarce in most areas and very abundant in a small proportion (e.g. Taylor and Taylor 1979; Taylor 1986; Cammell *et al.* 1989).

Over a substantially larger area, Brewer and Gaston (2002) estimated the density of the holly leaf-miner *Phytomyza ilicis*, a specialist miner of the leaves of European holly *Ilex aquifolium*, at multiple sites across its predominantly European geographic range. They found that the species was absent from a high proportion of the trees sampled, occurred at low densities on most of the rest, and at high densities on only a very small number (Fig. 4.2). However, in common with many studies of intraspecific abundance distributions, it is not clear how representative the sampled areas were of the whole of the range of the leaf-miner—in purposely sampling both central and peripheral areas they may well have been somewhat biased, although integrating under an interpolated abundance surface generated for the whole geographic range gave much the same result. The *post hoc* demonstration of representativeness (in this case, a random sampling of areas, with respect to abundance) is not always straightforward, and studies at large scales have for practical reasons seldom been designed explicitly to address this problem.

Based on population size estimates for 50×50 km grid squares, Gregory *et al.* (1998) found that across Europe bird species were also rare in most areas and abundant in a few. Common species typically exhibited left-skewed distributions of

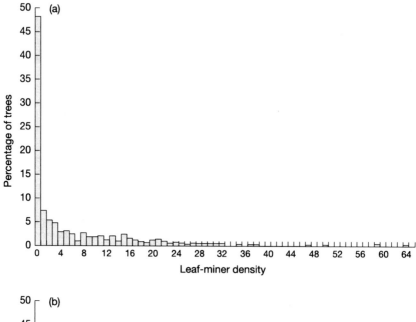

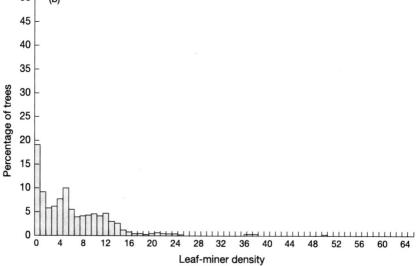

Fig. 4.2 Intraspecific abundance distributions for the holly leaf-miner *Phytomyza ilicis* across its natural geographic range (a) on host trees sampled and (b) based on an interpolated abundance surface across the range. Abundance is mines per 200 leaves sampled. From Brewer and Gaston (2002).

log-transformed abundance, and scarce or rare species exhibited right-skewed distributions under such a data transformation (although some of these species have ranges that extend beyond the bounds of the analysis). Because there are many more scarce or rare species than common ones, the frequency distributions thus tended on average to be right-skewed.

At a yet broader scale, most of the work that has been done on intraspecific abundance distributions for terrestrial species has concerned North American birds. Thus, Gaston (1994a), using smoothed abundance maps for wintering birds in North America (Root 1988a), observed that for each of eight species of wren (Troglodytidae) higher densities covered progressively smaller areas of their geographic ranges. Brown *et al.* (1995) conducted more formal analyses of the patterns of variation in the abundances of 90 species of passerine birds whose geographic ranges lie largely or entirely within eastern and central North America, using data from the Breeding Bird Survey. They found that the abundance of each species at the sites where it occurs varies from many 'cool spots' where it is rare, to a few 'hot spots' where it is orders of magnitude more abundant. Indeed, for 77 of the 90 species, more than half of the total number of individuals was recorded on less than a quarter of the survey routes where that species occurred (Fig. 4.3).

There has been much debate as to the best models for describing intraspecific abundance distributions. Various ones have been championed, but it is clear that, broadly speaking, none exhibits the flexibility of forms exhibited by real data. The negative binomial distribution is the most widely employed, with the probability of observing x individuals in a sampling unit (P_x) taking the form

$$P_x = \left(1 + \frac{\mu}{k}\right)^{-k} \frac{(k+x-1)!}{x!(k-1)!} \left(\frac{\mu}{\mu+k}\right)^x \qquad (4.1)$$

where μ is the mean number of individuals per sampling unit, and k is a clumping parameter. Small values of k represent strong aggregation moving towards a random distribution as values of k increase. A large number of possible causal derivations of the negative binomial have been identified, most based on the compounding of random processes (e.g. Boswell and Patil 1970; Taylor 1984).

This model has a relationship between the variance (σ^2) and μ (across unoccupied and occupied sites) attained by a species (with data points representing either mean and variance determined across different sets of sites at a given time or across the same set of sites at different times) that takes the form

$$\sigma^2 = \mu + \frac{\mu^2}{k} \qquad (4.2)$$

(Routledge and Swartz 1991; Perry and Woiwod 1992; Gaston and McArdle 1994). In other words, σ^2 is a positive curvilinear function of μ. Much of the discussion of the

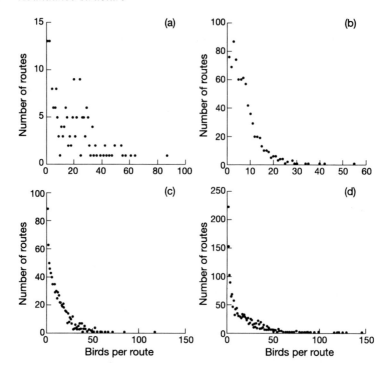

Fig. 4.3 Frequency distributions of abundance of four species of North American passerine birds among Breeding Bird Survey census routes; (a) scissor-tailed flycatcher *Muscivora forficata*, (b) Carolina chickadee *Parus carolinensis*, (c) Carolina wren *Thryothorus ludovicianus*, and (d) red-eyed vireo *Vireo olivaceus*. From Brown *et al.* (1995).

pattern of spatial variation in the abundances of species has concerned such variance–mean relationships, rather than the form of individual frequency distributions.

The utility of the negative binomial model is compromised by the fact that k is density dependent (although this trend has predominantly been investigated at small spatial scales; Taylor *et al.* 1978, 1979; Nachman 1981; Perry and Taylor 1985, 1986; Shorrocks and Rosewell 1986; Feng *et al.* 1993)[3]. Indeed, typically, when they are low the densities of a species exhibit Poisson-type distributions—where

$$P_x = e^{-\mu} \frac{\mu^x}{x!}$$

gives the probability of observing x individuals in a sampling unit—and pass through negative binomial-type to lognormal-type distributions—for a lognormal

$$P_x = \frac{1}{x(\sigma^2 \, 2\pi)^{1/2}} \exp\left(\frac{(\log x - \mu)^2}{2\sigma^2}\right)$$

—at high densities (Perry and Taylor 1985). Arguably, such variation in shape is captured by the relationship formalized as Taylor's power law (Taylor 1961):

$$\sigma^2 = \alpha\mu^\beta \tag{4.3}$$

where α and β are constants.

Taylor's power law seems to fit reasonably well for abundance data from a range of spatial scales (Fig. 4.4; e.g. Taylor *et al.* 1978, 1979, 1980; Perry 1981; Maurer 1994), when low density values are removed, which tend to track the Poisson line $\sigma^2 = \mu$ (graphically illustrated by Taylor and Woiwod 1982; Taylor 1984, 1986). However, again, there has been much debate on the matter (Routledge and Swartz 1991; Perry and Woiwod 1992; McArdle and Gaston 1995). Crucial to the argument is the role of sampling error in estimates of variability, because this may have a substantial influence on the shape of the mean–variance relationship that is observed, particularly because sampling error is likely to have a disproportionate influence at low mean densities. Indeed, the distinction between observed and actual densities should be remembered throughout this chapter; empirical data are seldom adequate to separate observed from actual densities, but theory deals with actual densities. Methods of removing the effects of sampling error to isolate the real underlying relationship between mean and variance are receiving growing interest, and seem to

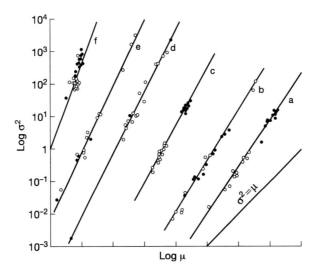

Fig. 4.4 Relationships between the mean and variance of densities of bird species at sites across Britain, with Taylor's power law fitted, for (a) woodcock *Scolopax rusticola*, (b) stonechat *Saxicola torquata*, (c) tawny owl *Strix aluco*, (d) rook *Corvus frugilegus*, (e) sand martin *Riparia riparia*, and (f) linnet *Carduelis cannabina*. Closed circles for woodland sites, open circles for farmland sites. From Taylor *et al.* (1980).

reveal that these may be somewhat different (McArdle and Gaston 1995; Stewart-Oaten *et al.* 1995).

Most exploration of the form of intraspecific abundance distributions has employed statistical rather than ecological models. Nonetheless, a number of ecological models, rooted in a wide range of population processes, have proven to be capable of predicting the form of Taylor's power law, including models based on ideal free distributions, behavioural dynamics, demographic stochasticity, and chaotic population dynamics (e.g. Anderson *et al.* 1982; Binns 1986; Gillis *et al.* 1986; Perry 1988, 1994; statistical artefacts may also play a role—Soberón and Loevinsohn 1987). In short, a wide range of processes can give rise to what is observed. For example Perry (1988) presents two population dynamic simulation models (as well as a stepping-stone model) that can produce realistic variance–mean relationships. In the first, in any spatial unit there is assumed to be a maximum population size above which emigration of additional individuals occurs, with this population size being lognormally distributed across spatial units, most emigrants from a unit dying, immigration of the remaining individuals to each unit being at a random rate, and population growth rates within a spatial unit being randomly sampled from a lognormal distribution. In the second model, density dependence within a spatial unit was assumed to operate at all densities, following a discrete-time difference equation analogue of the logistic, and with the maximum population size attainable either varying spatially according to a lognormal distribution and remaining constant through time or with some temporal variation also. Such models seem likely to capture the essence of the population dynamics of a wide diversity of species.

A problem that has persistently plagued debate about the validity of models describing intraspecific abundance distributions is that any coincidence between the distributions predicted by a model and those observed provides no evidence that this model bears any relationship to the processes actually determining the observed patterns (though the lack of any coincidence is suggestive that the processes embodied in the model are unlikely to be the correct ones). This highlights the more general issue that the similarity of two patterns, whether observed or modelled, does not necessarily mean that they have been generated by the same process. The fact that several plausible ecological processes can generate patterns in the frequency of local abundances across a species range that are closely akin to those that are actually observed suggests that different processes may give rise to observed patterns in different cases.

4.3 Intraspecific abundance–range size relationships

A feature common to many statistical functions that have been used to describe the spatial distribution of the individuals of a species is that as total population size increases so does area of occupancy, and that this happens in such a fashion that occupancy increases with increases in local density (Wright 1991; Gaston *et al.* 1997a, 1998c;

Hartley 1998). Thus, for example, the negative binomial distribution yields a positive curvilinear relationship, such that

$$p = 1 - \left(1 + \frac{\mu}{k}\right)^{-k} \tag{4.4}$$

where p is the proportion of sites or areas that are occupied by the species and μ is its mean abundance across all the sites or areas. As k increases a species would exhibit increasing occupancy for a given mean abundance.

Positive intraspecific relationships between abundance and occupancy, or evidence suggestive of such relationships, have been documented by studies of a number of species (Figs 4.5 and 4.6). These include investigations of plants, butterflies, moths, fish, and birds[4]. Such relationships are, however, substantially less consistent than interspecific abundance–occupancy relationships (Section 3.5.9), and some analyses have failed to document them in at least some instances (Fig. 4.6; Ambrose 1994; Marshall and Frank 1994; Swain and Morin 1996; Boecken and Shachak 1998; Donald and Fuller 1998; Gaston *et al.* 1998d, 1999a)[5]. Unfortunately, the bulk of analyses of intraspecific patterns are not strict equivalents of those for interspecific abundance–occupancy relationships. Rather, the majority, concern estimates of mean abundance and occupancy across the same sets of sites for different periods, instead of different sets of sites at the same time, or approximately so (but see Venier and Fahrig 1998). In other words, they document the fact that when a species is locally abundant it tends to be relatively widespread, and that when it is locally rare it tends to be more restricted in its occurrence. Despite this, it seems reasonable to assume that intraspecific equivalents of interspecific relationships are commonplace, and thus that where a species is locally abundant it is relatively widespread, and where it is locally rare its distribution is more restricted.

The fit of statistical models to intraspecific abundance–occupancy relationships remains to be explored in any detail for data at geographic, let alone whole range, scales. Nonetheless, in addition to negative binomial models, other abundance–occupancy models have been proposed largely as empirical descriptors of observed patterns, particularly at smaller spatial scales. The first of these was originally suggested by Nachman (1981), and takes the form

$$p = 1 - e^{-\alpha\mu^{\beta}} \tag{4.5}$$

or

$$\log\left(\log\left(\frac{1}{1-p}\right)\right) - \log\alpha + \beta\log\mu \tag{4.6}$$

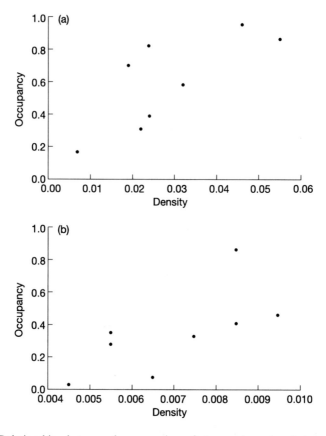

Fig. 4.5 Relationships between the proportion of streams in major drainages of North Carolina and Virginia occupied and densities in those streams (number/m²) for (a) brook trout *Salvelinus fontinalis* ($r_s = 0.755$, $n = 8$, $p < 0.05$) and (b) brown trout *Salmo trutta* ($r_s = 0.819$, $n = 8$, $p < 0.05$). From data in Flebbe (1994).

where *a* and *b* are two positive parameters. This model has principally been used to predict the densities on crops of pest species from data on their occupancy of sampling units (plants, tillers, leaves, etc.; Nachman 1981, 1984; Kuno 1986, 1991; Ward *et al.* 1986; Ekbom 1987; Perry 1987; Hepworth and MacFarlane 1992; Feng *et al.* 1993). It predicts a relationship between *m* and s^2 of the form of Taylor's power law (Eq. (4.3)).

Albeit in an interspecific context, Hanski and Gyllenberg (1997) use a logistic model to describe the relationship between abundance and occupancy, such that

$$p = \frac{1}{1 + e^{-\alpha - \beta \log \mu}} \tag{4.7}$$

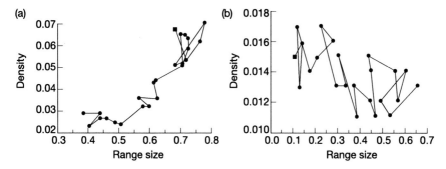

Fig. 4.6 The intraspecific relationship between abundance (territories/ha when present) and range size (proportion of sites occupied) for (a) tree sparrow *Passer montanus* and (b) sparrowhawk *Accipiter nisus* on farmland Common Birds Census sites in the period 1968–91. The points are joined in temporal sequence, with 1968 indicated by a square point. In years when they occurred on a higher proportion of sites, tree sparrows were more abundant on those sites than in years when they occurred on fewer sites, and sparrowhawks were less abundant. From Gaston *et al.* (1999).

or

$$p = \frac{\alpha\mu^{\beta}}{1 + \alpha\mu^{\beta}} \tag{4.8}$$

or

$$\log\frac{p}{1-p} = \alpha + \beta\log\mu \tag{4.9}$$

and

$$\mu = \left(-\frac{1/p-1}{\alpha}\right)^{-1/\beta} \tag{4.10}$$

where α and β are two positive parameters. For this model, the relationship between the spatial variance in abundance and mean abundance takes the form

$$\sigma^2 = \alpha\mu^{\beta}(1 + \alpha\mu^{\beta}) \tag{4.11}$$

which is larger than the variance in the Taylor's power model, suggesting that the Hanski–Gyllenberg model is appropriate to describe patterns for species having stronger aggregation than that under the Nachman model.

Again in an interspecific context, Leitner and Rosenzweig (1997) suggest a simple power model for abundance-range size relationships, such that

$$p = \alpha\mu^{\beta} \tag{4.12}$$

or

$$\log p = \log\alpha + \beta\log\mu \tag{4.13}$$

where α is positive and β is a scale parameter. This model has a variance–mean relationship of the form

$$\sigma^2 = \alpha\mu^{\beta}(1 - \alpha\mu^{\beta}) \tag{4.14}$$

with $\alpha\mu^{\beta} < 1$. Compared with the Nachman model, the power model is suitable for species of less aggregated or regular distribution.

When the spread of mean abundance values becomes moderate to large, the abundance–occupancy relationships that are predicted by several of these models become quite similar for all but the rarest species. This means that individual species need only show limited commonality in their aggregative behaviour for an interspecific abundance–occupancy relationship of the kind typically observed (Section 3.5) to be found.

A very general model of abundance–range size relationships (both intraspecific and interspecific), of which many others (Poisson, negative binomial, Nachman) are special cases, takes the form:

$$p = 1 - \left(1 + \frac{\alpha\mu^{\beta}}{k}\right)^{-k} \tag{4.15}$$

or

$$\mu = \frac{k}{\alpha}\left[(1 - p)^{-1/k} - 1\right]^{1/\beta} \tag{4.16}$$

where α is a positive parameter, β is a scale parameter, and k is a real parameter defined in the domain of $(-\infty, -\alpha\mu^{\beta})$ or $(0, +\infty)$ (He *et al.* 2002). For a given overall number of individuals and k greater than zero, as k increases occupancy increases. For a given overall number of individuals and k less than zero, occupancy declines as k assumes progressively greater negative values. When $k \rightarrow \pm\infty$ and $\alpha = \beta = 1$, this is the Poisson model, when $\alpha = \beta = 1$ it is the negative binomial model, when $k \rightarrow \pm\infty$ it is the Nachman model, when $k = 1$ it is the Hanski–Gyllenberg model, and when $k = -1$ it is the power model (He *et al.* 2002).

Combining models (4.2) and (4.4), and thereby linking mean–variance (in abundance) and abundance–range size relationships, gives the unified general model:

$$p = 1 - \left(\frac{\mu}{\sigma^2}\right)^{(\mu^2/(\sigma^2-\mu))} \tag{4.17}$$

where $\sigma^2 \neq \mu$ but can infinitely approach μ resulting in $p = 1 - e^{-\mu}$, which is occupancy for the Poisson distribution (He and Gaston ms). For empirical data this model can yield good predictions of occupancy based on mean and variance in abundance (Fig. 4.7).

This linking of two major relationships in areography may not necessarily serve greatly to advance understanding of the determinants of these patterns, in that it explains one descriptor of the spatial distribution of individuals (the mean–variance relationship) in terms of another (the abundance–range size relationship; Gaston *et al.* 1998c). Importantly, however, it highlights the fact that the mechanisms giving rise to one pattern will also give rise to the other. Thus, if realistic variance–mean relationships can be generated from models of ideal free distributions, behavioural dynamics, demographic stochasticity, and chaotic population dynamics, for example, so can abundance–range size relationships. The reverse is also true. With the exception of that based on phylogenetic non-independence, there are intraspecific equivalents of all of the hypotheses that have been proposed to explain interspecific relationships between abundance and range size, including those based on sampling artefacts, range position, resource breadth, resource availability, habitat selection, metapopulation dynamics, and vital rates (Section 3.5.9; Gaston *et al.* 1997a, 2000). These mechanisms should also be capable of predicting variance–mean relationships. The limited overlap between these two sets of hypotheses must, therefore, be an artefact of the development of the literature from which they derive, rather than symbolic of any mechanistic distinction. The fact that many population dynamic processes can give rise to these relationships should come as no surprise. If they are indeed quite general patterns then they should be produced by the full range of processes that dominate the population dynamics of different kinds of organisms.

In the context of abundance–range size relationships, in part because of the significance for conservation and management issues, particular attention has been paid to a metapopulation dynamic mechanism in which species dispersal is assumed to be a fundamental process in maintaining local populations in different habitat patches (Hanski 1991a; Hanski and Gilpin 1997). From the prediction of the unified model (4.17), such dynamics should also lead to a positive variance–mean relationship. Indeed, the variance–mean pattern is inevitably produced by metapopulation dynamics, through immigration of individuals from high density sites to lower density or vacant ones (the rescue effect; Hanski 1991a; Hanski and Gilpin 1997). This effect increases the chance of colonization and reduces spatial variation, conforming to the prediction of model 4.17. These theoretical results are not only supported by empirical field experiments (Gonzalez *et al.* 1998; Kruess and Tscharntke 1994) but

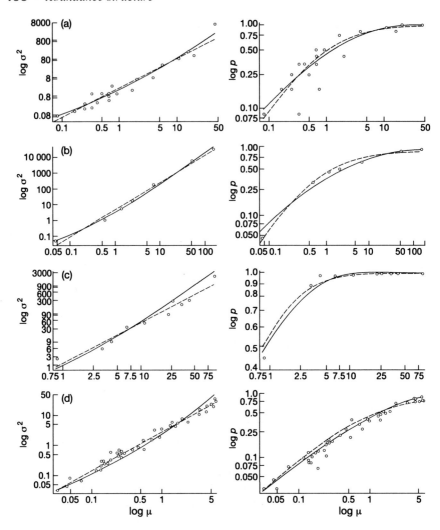

Fig. 4.7 Intraspecific variance–mean and abundance–occupancy relationships at a variety of spatial scales for (*a*) striped ambrosia beetle *Trypodendron lineatum*, (*b*) the aphid *Acyrthosiphon pisum*, (*c*) the tick *Ixodes ricinus*, and (*d*) the chrysomelid *Altica oleracea*. The dashed lines/curves are the power model $\sigma^2 = \alpha\mu^\beta$ and the occupancy model $p = 1 - (\mu/\sigma^2)^{(\mu^2/\sigma^2 - \mu)}$ fitted to the respective data. The solid curves are the predictions of $p = 1 - e^{-\mu}$ given by the opposite patterns. From He and Gaston (ms).

are also demonstrated by stochastic cellular automaton simulations, which show that spatial aggregation is an unavoidable outcome of poor colonization and dispersal ability, from which the predicted negative variance–occupancy correlation will result (Tilman *et al.* 1997).

It is well known in ecology that, all else being equal, the greater the temporal variation of a population the greater the likelihood of extinction (Leigh 1981; Goodman 1987). However, how spatial variation affects population persistence is poorly understood. According to model 4.17, for a given level of abundance high spatial variability is associated with a small range size, making a population more susceptible to environmental change and habitat loss and therefore increasing its risk of extinction. This is supported by much empirical evidence showing that landscape fragmentation or environmental stochasticity, which promotes spatial variability, results in lower occupancy (i.e. 'shallow incidence function'; Hanski 1992) and is a possible cause of population extirpation (Pimm 1991; Kruess and Tscharntke 1994; Gonzalez *et al.* 1998). Such a negative impact of greater spatial variability on species persistence may again be understood in terms of the rescue effect, which both reduces the likelihood of local extinction and lowers spatial variation. Rare species are typically found to be less aggregated than common species, which, sampling artefacts aside, may reflect the fact that low aggregation is the means by which they persist.

4.4 Environmental gradients and response curves

Discussion of the abundance structure of the geographic ranges of species in this chapter has, thus far, made no reference to where different levels of abundance occur within the range. For example that local abundances are distributed in, say, a negative binomial fashion conveys nothing about whether areas of high abundances occur close together or far apart. The most common way in which this has been addressed has been the partial response of asking, without any necessary explicit reference to the spatial relations of sample points, how the abundances of species respond to environmental gradients.

Spatial variation in the environment is, of course, complex. Three different kinds of environmental gradients have been distinguished (Austin *et al.* 1984): (i) indirect or complex gradients—the influence of these on growth is indirect (e.g. altitude, aspect), in that it is other factors that covary with these that have the direct influence; (ii) direct gradients—the influence of these on growth is direct (e.g. pH and temperature in the case of plants); and (iii) resource gradients—these involve factors that are directly used as resources by the species (e.g. nitrogen in the case of plants). Across such gradients, the abundance of a species is generally held to peak at some intermediate environmental condition and to decline in both directions away from this point; the presumption being that species are progressively less fitted to life at extremes of an environmental gradient (and hence that the distributions of species tend to be limited by abiotic factors). The particular pattern of change is commonly known as a response curve. Most attention has perhaps been paid to such curves with regard to their match (or otherwise) with the assumptions of ordination techniques, and therefore the suitability of these techniques for certain kinds of statistical analyses in ecology, rather than for their inherent interest (Austin 1976).

Because of the greater availability and ease of acquisition of appropriate data, most investigations of response curves have been for plants. In general, when determined across the fullest breadth of environmental conditions under which a species occurs such curves tend to be unimodal, although they are commonly also skewed (Austin 1976; Westman 1980; Silander and Antonovics 1982; Austin *et al.* 1985; Wilson and Keddy 1988; Huntley *et al.* 1989; Minchin 1989; De'Ath 1999); the direction of the skew may reflect the position occupied by a species on an environmental gradient, with skews increasing away from the centre of the gradient and in the direction of the centre (Austin and Gaywood 1994; Austin *et al.* 1994). Response curves may also exhibit no peaks, and may indeed appear to be approximately linear, although in some cases this may simply be because a narrower spread of conditions has been considered and thus only part of the full response is being sampled (as will typically occur in just part of the geographic distribution of a species; Hansen 1989; Odland *et al.* 1995).

Examples of response curves for animal species appear at first sight to be scant, although where they have explicitly been explored similar conclusions seem to pertain as for plants (De'Ath 1999). However, a number of analyses have sought to determine the relationships between the abundance of an animal species in different places and variables that typify the environments of those areas (Fig. 4.8; e.g. Rogers and Randolph 1991; Kauhala 1995; Green 1996; Bustamante 1997; Jarvis and Robertson 1999; Brewer and Gaston 2003)[6]; some of the more recent work has been

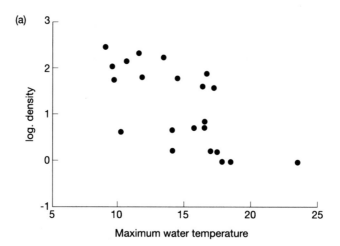

Fig. 4.8 Relationships between the abundances of species in different areas and the associated environmental conditions: (a) maximum water temperature (°C) and density of the freshwater sculpin *Cottus nozawae*, (b) mean annual normalized difference vegetation index (NDVI) and the mean daily trap catch of the tsetse fly *Glossina palpalis*, and (c) land productivity index and mean distance to nearest neighbour (km) of the sparrowhawk *Accipiter nisus*. From Newton *et al.* (1977), Rogers and Randolph (1991), and Yagami and Goto (2000).

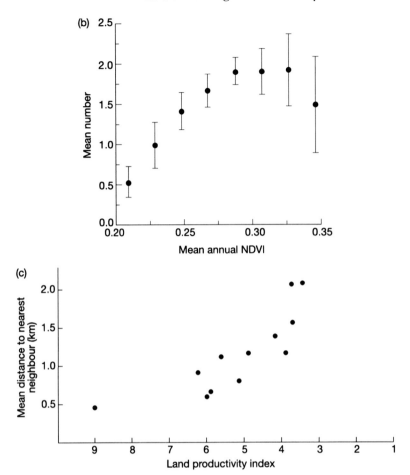

Fig. 4.8 *(continued)*

particularly concerned with environmental gradients associated with urbanization or pollution (e.g. McDonnell and Pickett 1990; Ruohomäki *et al.* 1996; Denys and Schmidt 1998). The search for such relationships is implicit in many of the attempts that have been made to model the distributions of species in terms of prevailing environmental factors (e.g. Section 2.2.2). Here, attention has focused principally on whether significant linear relationships exist or not (an observation made nearly 20 years ago; Meents *et al.* 1983). It remains an interesting question as to the proportion of non-linear relationships between abundances and environmental variables that have been encountered and not recognized as such. Certainly where they have been sought such relationships have commonly been found (e.g. Meents *et al.* 1983). Consideration of the fit of more complex terms might be considered as a routine check—although not necessarily of itself sufficient—on such matters.

For both plant and animal species, response curves have regularly been generated based on frequency of occurrence rather than abundances (e.g. Lawton *et al.* 1987; Hansen 1989; Westman 1991; Eyre *et al.* 1992a, 1992b, 1993; Austin *et al.* 1994; Austin and Gaywood 1994). Here, the response to environmental variation is the proportion of sites or areas that share a given range of environmental conditions at which the species occurs. Austin (1987) performed such an analysis for eucalypt species in south-eastern New South Wales, examining the shape and distribution of species response curves along a temperature gradient for 750 sites, having carefully stratified a larger data set to control other forms of environmental heterogeneity. He found that bell-shaped curves were not universal, and that positive- or right-skewed curves were characteristic of major canopy species. Indeed, the patterns resulting from such analyses seem to be broadly the same as those based on response curves for species abundances.

As already mentioned, the predominant interpretation of the shape of response curves is that it reflects the suitability of environmental conditions for a species. This is arguably a rather superficial interpretation, in as much as it demands a tighter link between abundance and environment than might be reasonable (see Section 4.5.1), although it probably holds as a rough and ready generalization. This would essentially connect response curves to population dynamic models of intraspecific abundance distributions through the frequency of different sets of environmental conditions and hence the frequency of local carrying capacities of different magnitudes. This argues that these frequencies should be modelled as modal, and probably skewed, functions.

The general similarity of response curves based on abundances and on occurrences (presence/absence) hints at a possible fruitful line of investigation centred on exploring the relationship between the abundance and occupancy of a species, where data points are each determined for a different set of environmental conditions. Do species exhibit higher levels of occupancy for those environmental conditions under which they are more abundant? I am not aware of studies of this kind.

4.5 Abundance profiles

4.5.1 Patterns

Across the entire geographic range of a species it is unlikely that there will be any simple pattern of variation in an environmental variable, and these patterns are likely to vary between variables; this is true even across latitudinal gradients, although these are commonly characterized as having very simple patterns of environmental variation. Thus, of themselves, response curves still provide only limited insight into the spatial relations of different levels of abundance across the distribution of a species.

In the context of geographic ranges, consideration of these spatial relations has largely concerned the question of how the abundance of a species changes toward

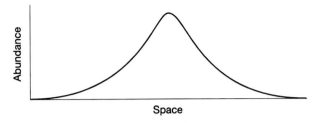

Fig. 4.9   Visualization of a hump-shaped abundance structure to a geographic range.

the range limit or periphery. It has long been widely accepted that, to a first approximation, the local abundance of a species can be regarded as tending to peak towards the centre of its geographic range and to decline towards the periphery (Shelford 1911; Kendeigh 1974; Hengeveld *et al.* 1979; Lawton 1993; Safriel *et al.* 1994; Hochberg and Ives 1999)[7]. Indeed, such a pattern has already been alluded to several times in this book. That is, the abundance structure of a geographic range can, at least in many cases, be visualized as essentially hump-shaped (Fig. 4.9).

Direct empirical support for a decline in the abundance of species towards the edges of their geographic ranges arises from studies of a diversity of taxa (plants, insects, molluscs, fish, birds, mammals)[8]. These studies take a variety of forms, including visual inspection of maps of raw or raw and interpolated abundance data (e.g. Fig. 4.10), analyses of relationships between local abundances and position along transects across geographic ranges, analyses of spatial autocorrelation functions, and interspecific relationships between local abundance and proximity to range edge. For example:

1. Gyrinus *beetles*—Svensson (1992) studied the abundance patterns of three species of *Gyrinus* beetles in northern Europe. All three have broad habitat ranges, and opportunistically exploit a variety of kinds of water bodies, with *G. natator* and *G. substriatus* having a northern limit in Sweden, and *G. opacus* having a southern limit there. Data collected by sampling at multiple localities revealed that the abundance of all three in temporary pools increased with distance from their respective range limits (Fig. 4.11).

2. *Insectivorous passerines*—Tellería and Santos (1993) analysed patterns in the abundances of each of six species of foliage insectivorous passerines in 58 woodlands along a 850 km belt crossing the Iberian peninsula. Five of the six species showed significant negative correlations between their abundances and the distance of forests to the north (Fig. 4.12). The authors argued that this supported a decline in abundance from range centres to periphery, because this region is on the southern most borders of the Palearctic, and the species have a core area in central-east Europe.

3. *North American birds*—Brown *et al.* (1995) used data from the Breeding Bird Survey to evaluate the geographic patterns of abundance of species across

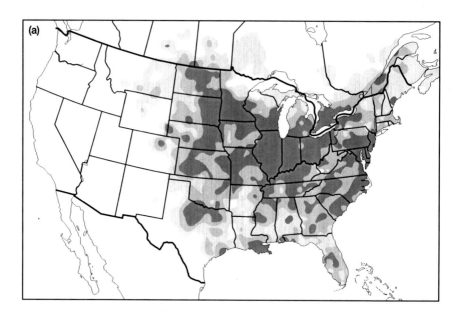

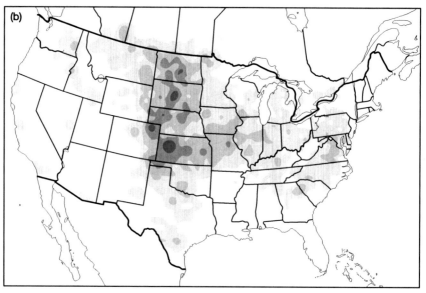

Fig. 4.10 Spatial variation across North America in the abundance of (a) common grackle *Quiscalus quiscula*, (b) grasshopper sparrow *Ammodramus savannarum*, (c) dickcissel *Spiza americana*, and (d) northern cardinal *Cardinalis cardinalis*. Higher levels of abundance are represented by darker shading, in the sequence < 5, 5–20, 20–50 and > 50 birds detected per sample route per year. From Price *et al.* (1995).

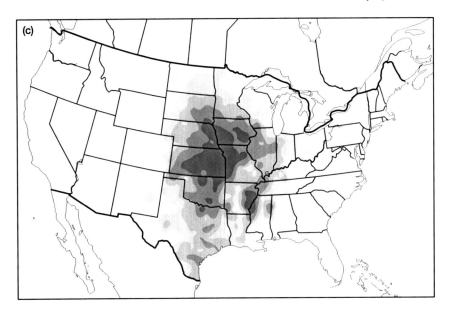

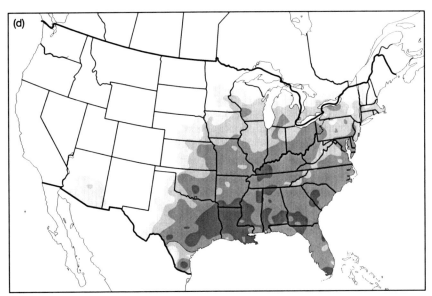

Fig. 4.10 *(continued)*

North America. Spatial autocorrelation analyses of local abundances generated correlograms (a plot of spatial autocorrelation [Moran's I, which is functionally equivalent to a correlation coefficient] against the distance between points) that showed two peaks of positive values indicating that sites are similar in the

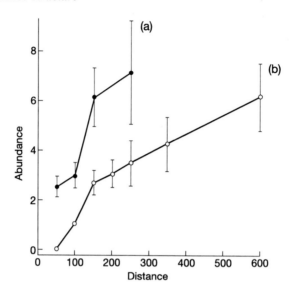

Fig. 4.11 Abundance (mean±SE) of the water beetles (a) *Gyrinus substriatus* and (b) *G. opacus* in temporary pools as a function of distance (km) from their northern and southern range limits, respectively. From Svensson (1992).

abundance of a species (Fig. 4.13; Villard and Maurer 1996 present related analyses based on semivariograms, plots of semivariance against the distance between points). One of these peaks was at short distances, corresponding to sample sites throughout the geographic range, and the other at maximum distances, corresponding to sites at opposite edges of the range. They interpreted these to reflect the fact that sites close together tend to have similar abundances, no matter whether those abundances be high or low, and that sites very far apart tend to be at the periphery on opposite sides of a species geographic range and to share low abundances. To some extent, this interpretation was supported by looking at mean abundance as a function of distance from the edge of the range (Fig. 4.14).

Each of the kinds of evidence that have been proffered in support of a decline in the abundances of species towards their range limits may suffer from one or more important limitations:

1. *Visual interpretations of abundance surfaces generated for all or part of a species' geographic range*—these are problematic on the grounds that the human visual system is adept at seeing in images simple patterns that may, in fact, not be present (at least in statistical terms), particularly when there is some preconception as to what will be found. Indeed, it is salutary to present several individuals with a set of such maps (e.g. Fig. 4.10), and to ask them independently to describe the patterns they observe. Variance in the responses can be profound.

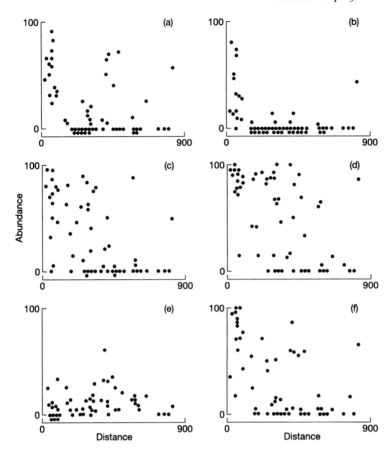

Fig. 4.12 Abundance in 58 forests as a function of distance (km) along a gradient from the Spanish–French boundary (0 km) to the southernmost tip of Spain of six insectivorous passerines: (a) blackcap *Sylvia atricapilla*, (b) chiffchaff *Phylloscopus collybita*, (c) firecrest *Regulus ignicapillus*, (d) robin *Erithacus rubecula*, (e) long-tailed tit *Aegithalos caudatus*, and (f) wren *Troglodytes troglodytes*. From Tellería and Santos (1993).

2. *Analyses of abundance variation for only parts of species' geographic ranges*— these may be misleading of wider patterns, if those patterns are complex or variable between different parts of the range. In some cases transects do not reflect well the full range of environmental conditions that a species may exploit, leading to biased results (which relates back to comments about response curves; Section 4.4). Thus, in a study of the pattern of abundance of the annual weed prickly lettuce *Lactuca serriola* along the verges of a motorway in England, colonies did not become smaller or sparser towards its northern or southern distribution limits, but in a wider survey there was a slight decline in the frequency

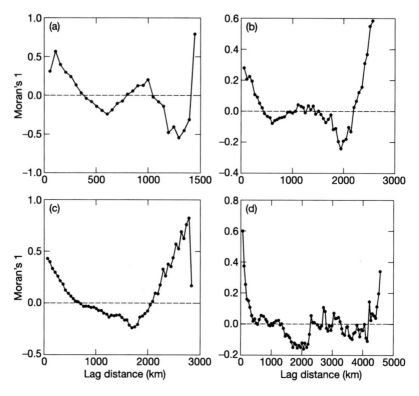

Fig. 4.13 Spatial autocorrelation of abundance of four species of North American passerine birds among Breeding Bird Survey census routes: (a) scissor-tailed flycatcher *Muscivora forficata*, (b) Carolina chickadee *Parus carolinensis*, (c) Carolina wren *Thryothorus ludovicianus*, and (d) red-eyed vireo *Vireo olivaceus*. From Brown *et al.* (1995).

of colonies towards the limit (Prince *et al.* 1985). Often transects are chosen so as to exhibit simple strong environmental gradients, which may particularly bias results.

3. *Spatial autocorrelograms of abundances for sites across all or part of a species' geographic range*—whilst if such plots exhibit a U-shape (with higher autocorrelations at short and long lag distances) this might be generated by local abundances towards the centre of a range being high and those towards the limits being low (with high autocorrelation at short distances reflecting the similarity in abundances of sites close together, and the high autocorrelation at long distances reflecting the similarly low abundances at the opposing edges of the range), this pattern is only suggestive. Other forms of spatial variation in local abundance may yield the same result. Indeed, Brown *et al.* (1995) give four examples of bird distributions that do show this autocorrelation pattern using data from the

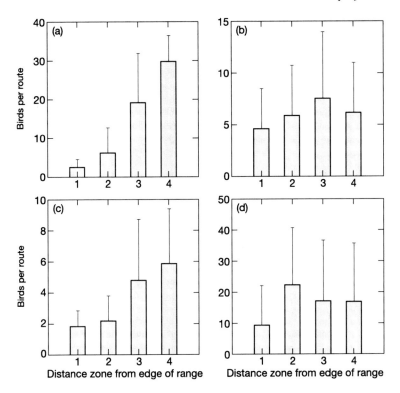

Fig. 4.14 Mean abundance per Breeding Bird Survey census route as a function of distance from the edge of the geographic range for four species of North American passerine birds: (a) scissor-tailed flycatcher *Muscivora forficata*, (b) Carolina chickadee *Parus carolinensis*, (c) Carolina wren *Thryothorus ludovicianus*, and (d) red-eyed vireo *Vireo olivaceus*. Data are plotted in quartiles of distance intervals, scaled from the edge to the centre of the range. Sites whose closest edge of range was a coastline were not included. From Brown *et al.* (1995).

North American Breeding Bird Survey. However, reference to abundance maps for these species generated from the same data source (Price *et al.* 1995) reveal that, arguably, none of them have a central abundance peak. Most notably, the map for the Carolina wren *Thryothorus ludovicianus* is extremely asymmetric with peak abundances toward the south-east coast of the United States, yet it's autocorrelation profile is plainly U-shaped, probably because the shape of the range means that the most northerly and westerly parts (which both have low densities of birds) are the most distantly separated.

4. *Interspecific patterns showing that species closer to their range limits have lower abundances*—these may indeed arise because abundances decline toward range limits, but may equally arise because those species at the edges of their ranges have smaller geographic ranges (and species with smaller ranges have

lower densities; Section 3.5.9). Likewise, the absence of a correlation between the abundances of species and their proximity to range edges (e.g. Gillespie 2000) provides little evidence as to changes in local abundance across the ranges of individual species.

There are also a number of other problems in determining the pattern of abundances across geographic ranges. First, observed patterns of spatial variation in abundances may, in part, rest on the spatial resolution of analysis (Wiens 1989). Second, if the habitat patch size occupied is smaller toward range limits, but size of survey area remains constant and is larger than patch size at range limits, then density will be observed to decline toward range limits even if, at the scale of the habitat patch, density remains constant. Third, changes in occupancy may provide limited proxies for changes in local abundance. Carter and Prince (1988) have argued that for plants, whilst in general there is evidence for a decline in the frequency of populations (occupancy) towards range limits, there is little evidence to support the view that populations become smaller (see also Griggs 1914); many studies have reported that populations become more sporadic towards range limits (Fig. 4.15; e.g. Svensson 1992; Kauhala 1995; Thomas *et al.* 1998; Nantel and Gagnon 1999; García *et al.* 2000; Yagami and Goto 2000)[9]. However, plainly this is not universal. Carey *et al.* (1995) found a decline in the area, density and population size of a winter annual grass *Vulpia ciliata* at four sites from the centre to the edge of its range, and one might expect some similarity of geographic pattern in

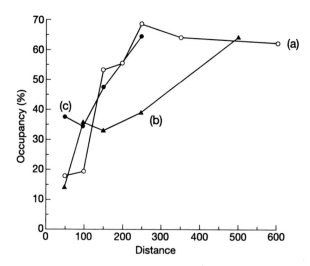

Fig. 4.15 Occupancy of sites by (a) *Gyrinus substriatus*, (b) *G. natator*, and (c) *G. opacus* as a function of distance (km) from their northern (a and b) and southern (c) range limits, respectively. From Svensson (1992).

abundance and occupancy based on the existence of intraspecific abundance–range size relationships (Section 4.3).

Whilst the majority of relevant studies that have been reported have claimed to find support for a decline in abundance toward range limits, there are a number of case studies which find conflicting patterns or less convincing evidence (e.g. Griggs 1914; Gilbert 1980; Rapoport 1982; Carter and Prince 1985a; Thomas *et al.* 1998; Blackburn *et al.* 1999)[10]. For example Griggs (1914) argued that for plants in the Sugar Grove region of Ohio, '... the species in which the individuals become scarcer and scarcer until it fails altogether is exceptional. In the majority the individuals are abundant in their respective stations up to the very edge of their ranges'. In a detailed analysis of birds in Britain (i.e. across only part of their geographic ranges), using five different measures of the proximity of areas to range limits, Blackburn *et al.* (1999) failed to find evidence that density declines towards these limits for most of the 32 species of passerines they examined (only 13 species showed significant relationships with any of the measures, and no more than seven species with any single measure). There are several methodological reasons why no such relationships might be forthcoming. However, it seems likely that their absence is genuine, although these species may become rarer towards their range limits in other senses (e.g. being more patchily distributed).

From first principles, there are a number of circumstances under which simple declines in abundance towards range limits seem unlikely to occur:

1. *Expanding populations*—for expanding populations, densities tend often to be relatively high close to the range limit and to be lower closer to the range centre (e.g. Caughley 1970; Baker 1985). This pattern is thought to result from a 'boom and bust' type dynamic, as densities build rapidly when a species first enters an area (which may be important for further expansion) and subsequently decline as the range front moves on and resources are depleted or populations of natural enemies respond.

2. *Unidirectional movement*—in environments where the diffusion or movement of individuals may be essentially unidirectional then abundance patterns may not show a central peak, but rather some systematic decline away from a peak at a range edge. This may occur, for example, in some marine and freshwater systems, where the flow of water moves individuals in one direction.

3. *Physical barriers*—where abrupt physical barriers disrupt environmental gradients without imposing any important larger-scale effects then abundances may suddenly drop when those barriers are reached. This circumstance has traditionally been considered to be rather scarce, but may actually be rather common, particularly at coasts.

In the extreme, of course, the simple characterization of the abundance structure of geographic ranges as hump-shaped cannot be correct. Even the most abundant species do not exhibit continuous distributions when viewed at fine resolutions. Rather, distributions are mosaic-like, with occupied areas interspersed with

unoccupied ones (Section 3.2). At the very least, a more appropriate general model
is one in which species near the edges of their ranges show uniformly low abundances, whilst those near the centres exhibit a wide range of abundances (Fig. 4.16;
Brown *et al.* 1995); the relationship between local abundance and distance from
range centre would thus be lower triangular (Enquist *et al.* 1995). Hengeveld (1990a)
suggests that similar distributions of densities occur at local scales as at regional ones
(as also intimated by various detailed population studies—Carey *et al.* 1995; but note
that smoothing functions will tend to produce the impression of such an effect). Thus
the abundance structures of the geographic ranges of species would be visualized as

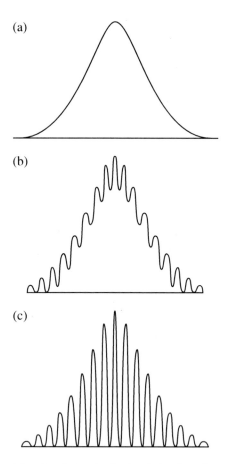

Fig. 4.16 Three models of variation in the local density of a species across a transect
passing from the edges through the centre of its geographic range: (a) simple unimodal
pattern, (b) unimodal pattern plus limited variance, and (c) unimodal pattern plus
high variance.

a series of abundance peaks, with the magnitudes of these peaks declining towards the range periphery. Pielou (1979, p.274) provides examples of species with disjunct distributions in which densities appear to decline toward the limits of each of the disjunct populations. The variety of possible spatial surfaces of abundances is, of course, constrained to fit what we already know about the form taken by intraspecific species–abundance distributions (Section 4.3), with many areas of low and only few of high abundance. But an almost infinite variety may fall within such constraints.

On balance, it seems unlikely that the abundance structures of the geographic ranges of the vast majority of species, even when mapped at moderate to coarse spatial resolutions, can be regarded as hump-shaped in anything but the vaguest of senses. Where abundance surfaces do exhibit a single, clear peak away from the range periphery, this peak will usually not be at the physical centre of the range. Most species will not exhibit a single peak, and many will have abrupt changes in their abundances towards their range limits.

4.5.2 Mechanisms

Several different models have been proposed that predict what form will be taken by the spatial distribution of levels of abundance across the geographic ranges of species. Here, I distinguish between two broad groups, based, respectively, on optimum response surfaces and on local population dynamics. These constitute different levels of explanation, rather than competing mechanisms, akin to the distinction drawn between abiotic and biotic factors on the one hand and population dynamics on the other when considering the determinants of geographic range limits (Chapter 2), and between the environmental constraint explanations for response curves and the population dynamic explanations for intraspecific abundance distributions.

Optimum response surface models

Perhaps foremost, as with response curves, the abundance structure of a geographic range has been seen as an optimum response surface, reflecting spatial variation in environmental conditions. Indeed, it has long been held that, with respect to the requirements of a particular species, environmental conditions deteriorate (become ecologically more marginal) from the centre towards the periphery of its range (e.g. Shelford 1911; Griggs 1914; Hutchins 1947; Chiang 1961; Huffaker and Messenger 1964; Gorodkov 1986a)[11]. If this is so, and particularly following from the above discussion of response curves along environmental gradients (Section 4.4), it would perhaps not be surprising if local abundances responded to such variation.

Two authors in particular have championed the view of the abundance structures of the geographic ranges of species as optimum response surfaces (although others have discussed more specific models; Williams 1988; Hall *et al.* 1992). First, Brown (1984, 1995) has argued that if (i) the abundance and distribution of a species are determined by combinations of many physical and biotic variables, and that spatial variation in population density reflects the probability density distribution of the required combinations of these variables, and (ii) some sets of environmental variables are distributed independently of each other, and environmental variation is spatially autocorrelated, then it follows that density should be highest at the centre of the range of a species and should decline towards the boundary.

In a similar vein, Hengeveld (1990a, b) regards the abundance structure of the geographic range of a species as an optimum response surface, reflecting gradual variation in the environment in all directions away from the peak. If, within certain limits, species responses are constant throughout their ranges, then where the environment is particularly favourable a species will attain a high density and where it is unfavourable it will attain a low one. As the environment changes in time, the surface will also change. Hengeveld (1990a) argues that, depending on the sensitivity of species to environmental conditions, the geographical intensity distribution of those conditions, and their temporal frequency distribution in each locality, geographic abundance structures of many different shapes can occur, including symmetrical and unimodal and skewed and multimodal.

The principal problems for optimum response surface models are that they assume that:

1. *There is a tight link between density and the favourableness of ecological conditions*—the use of population density as a measure of habitat quality has repeatedly been challenged (e.g. Van Horne 1983; Wiens 1989), and certainly there is no inevitability that such a link will exist. A decoupling seems most likely to occur in environments that are strongly seasonal, temporally unpredictable, or spatially patchy, and for species that are ecological generalists, long-lived, or have a high reproductive capacity. Some subset of these features characterizes a high proportion of species. Indeed, there is growing evidence for the widespread occurrence of sink–source population dynamics, with source populations exhibiting a local demographic surplus and sink populations a deficit (Pulliam 1988), and the possibility under some circumstances of densities becoming high in the low quality sinks; it is also becoming clear that these two kinds of populations can, in practice, be exceedingly difficult to distinguish using measures normally obtainable (Watkinson and Sutherland 1995; Dias 1996). Curiously, the decoupling of density and habitat quality can be increased by human activities that alter habitats in such a way that organisms cannot correctly evaluate habitat quality (Kokko and Sutherland 2001).

2. *The response of a species to environmental conditions is constant throughout its geographic range*—this ignores substantial evidence to the contrary (Section 2.2.7; for a review see Spicer and Gaston 1999). In particular, as geographic ranges have expanded there has been selection and adaptation to the different abiotic and biotic conditions experienced (Hewitt 2000). This counters some of the effects on density that would arise from changes to environmental conditions acting on a consistent environmental response.

3. *There is a moderate to high level of spatial autocorrelation in environmental variables*—many species may experience sharp changes in environmental conditions that generate limits to their ranges, without environmental conditions changing to that point in such a way as to cause a marked decline in abundance. Moreover, whilst environmental conditions may exhibit marked patterns of positive autocorrelation over long distances, these mask considerable complexities to which species densities may be responsive.

4. *The favourableness of ecological conditions declines towards range limits*—as Soulé (1973) observed, not all ecologically marginal populations are peripherally located, and not all peripheral populations are ecologically marginal. Of course, this need not undermine optimal response models *per se*, only the notion that they give rise to a central peak of abundance. Likewise, optimum response models have been criticized on the invalid assumption that environmental gradients are necessarily continuous in space.

5. *Physical processes are not sufficiently strong to impose abundance structures on geographic ranges*—this is not always so, though it probably is in the majority of cases. Indeed, Gaylord and Gaines (2000) have shown how ocean currents can generate, for example, an exponential decline in equilibrium adult abundance.

Minchin (1989) argues that optimum response surface models are based on a false analogy with statistical sampling theory (response curves based on abundances are not probability density distributions, samples near the mode are not necessarily more common than samples in the tails). The analogy with sampling theory is certainly a poor one, but this does not threaten the models as commonly expressed.

In addition, optimal response surface models require that only a relatively small number of factors determine variation in local abundances. Brown *et al.* (1995) demonstrated this using a simple model in which the niche of each species consists of multiple, independent factors that limit the fitness of individuals and thus abundance and distribution, and the effect of variation in each factor was characterized by a normal distribution. If abundance and distribution were determined by large numbers of factors, then species tended to be very rare where ever they occurred.

In sum, whilst optimum surface models may provide a useful heuristic for thinking about the structure of geographic ranges, they should not be taken too literally.

Local demographic models

In their simplest form, optimum response surface models make no explicit statement about the population dynamic behaviour underlying the abundance structure of geographic ranges, beyond assuming that as the environment changes local abundance and distribution are likely also to change in systematic ways. There have, however, been several attempts to develop demographic models of range structure. Here, four such models are outlined. These differ in basic construction, although they share some features in common. As they should, they also share features with population dynamic models postulated to explain the determination of geographic range limits (Section 2.3).

Basin model

MacCall (1990) developed the basin model for the abundance structure of the geographic ranges of species, treating this as an optimum response surface (see also Gibson 1994). Founded on density-dependent habitat selection, this is an explicitly dynamic model in which habitat suitability or *per capita* population growth rate is depicted graphically as increasing downward, and habitats can be described as a continuous topographic surface of geographic suitability that has the appearance of an irregular basin (Fig. 4.17). Under an Ideal Free Distribution, a population would fill such a basin just as would a liquid (the filled volume of the basin is proportional to the total population size). The topography of the basin will determine the observed abundance structure of the geographic range of the species. If habitats are most suitable towards the centre of the range, then densities will be greatest there also and a hump-shaped abundance structure will result. The intraspecific abundance distribution will be determined by the frequency of different depths. The density at any one location is proportional to the density-dependent reduction in realized suitability at that location.

MacCall went on to elaborate the basin model with regard to the geographic dynamics of fish populations, but in the present context we are only concerned with its essential features. Foremost, it requires that species exhibit density-dependent habitat selection, occupying more habitats when densities are high and fewer when they are low (driven through intraspecific competition; see also Section 3.5.9). Many undoubtedly do so (e.g. O'Connor 1987; Wiens 1989; MacCall 1990; Marshall and Frank 1995; for a review see Rosenzweig 1991). For example the number of nesting habitats used by wrens *Troglodytes troglodytes* each year in Britain is an increasing

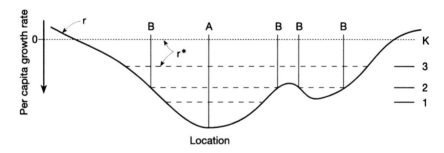

Fig. 4.17 The basin model, illustrated by a transect through a continuous geographic fitness topography. Realized suitability (per capita population growth rate), shown by the dashed line, is equal in all occupied habitats due to the ideal free distribution, and is given for three total population sizes (1, 2, and 3). A population size of carrying capacity (K) is reached when growth rate becomes zero. From MacCall (1990).

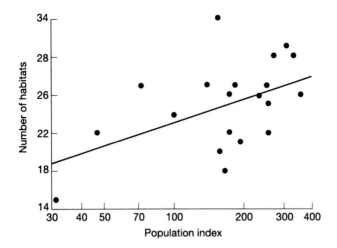

Fig. 4.18 Relationship between the number of habitats used by the wren *Troglodytes troglodytes* for nesting in each year and its population density on farmland (as measured by Common Birds Census index). From O'Connor (1987).

function of the population density of the species on farmland (Fig. 4.18; O'Connor 1987). Density-dependent habitat selection is, nonetheless, not universal, constraining the possible generality of the basin model. Perhaps more significantly, it seems unlikely that entire geographic ranges can be operating as Ideal Free Distributions once geographic ranges become of even moderate area. Tyler and Hargrove (1997), using simulation models, explore the likely validity of Ideal Free Distributions as predictors of the distributions of species over large areas. They found that predictions were poor when resources have a fractal rather than a random distribution, and when the scale at which distributions were measured was dissimilar to the maximum movement distance of foragers. This means that it is unlikely in many cases that Ideal Free Distributions will be appropriate descriptors of range wide dynamics, and therefore unlikely that the basin model will be appropriate. Additionally, they point out that whilst they may meet some of the assumptions and predictions of Ideal Free Distributions, studies that have marshalled support for the basin model have seldom, if ever, shown fit to the assumption that per capita gains or other measures of fitness were affected by increases in density.

Vital rates model
The vital rates model (Holt *et al.* 1997), discussed earlier as providing a possible explanation for positive interspecific abundance–occupancy relationships (Section 3.5.9), is essentially a model of the distribution of local abundances across the geographic ranges of species. It builds on ideas of Andrewartha and Birch (1954) and Lawton's (1993) diagrammatic descriptions of spatial variation in intrinsic population growth rates, r. The model assumes that r declines across an environmental

gradient from a central peak towards the extremes, and that there is a spatially invari-
ant density-independent death rate. Then, equilibrium local density varies directly
with *r*, peaking towards the centre of the environmental gradient.

Based on an environmental gradient, this model makes no explicit assumptions
about the spatial relations of different sites. As such, it can potentially predict a wide
range of abundance surfaces and is sufficiently general that it can account for substantial variation in intraspecific abundance distributions. However, such enormous
flexibility can also be seen as a weakness, making it potentially difficult to test the
predictions of the vital rates model.

A central presumption of the vital rates model is that intrinsic population growth
rates peak toward the centre of environmental gradients. The majority of work on
changes in population growth rates along such gradients has concerned laboratory
studies of insects, particularly species of agricultural importance, on temperature or
humidity gradients (Fig. 4.19; e.g. Lema and Herren 1985; Baldwin and Dingle
1986; Sands *et al.* 1986; Thaung and Collins 1986; Desmarchelier 1988; Acreman
and Dixon 1989)[12]. Here it appears that, in general, rates increase along the gradient
to a peak and then decline. Because the studies are largely concerned with identifying the point at which rates are maximized, they seldom consider the effects of conditions much beyond this point. However, it does seem that the pattern of change in
growth rates (and other performance measures) may be strongly asymmetric about
this peak, increasing steadily from lower temperatures, but decreasing dramatically
towards high ones (see Logan *et al.* 1976; Huey and Berrigan 2001).

Turning to responses to simultaneous changes in two environmental gradients, in
a classic set of laboratory experiments Birch (1953) examined variation in population growth rates of the two grain beetles *Sitophilus oryzae* and *Rhizopertha*

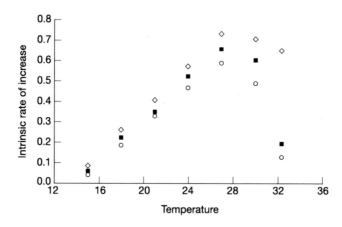

Fig. 4.19 Intrinsic rate of increase (per week) of the stored product beetle *Sitophilus oryzae*
at different temperatures (°C). Circles, 11 per cent moisture content of grain; squares, 12.5
per cent; and diamonds, 14 per cent. From data in Desmarchelier (1988).

dominica in relation to temperature and moisture. Population growth rate was expressed in terms of the finite rate of increase λ, where $\lambda = \text{antilog}_e\ r$. Across the range of environments explored, both species had their highest rates of increase at intermediate temperatures and at the highest moisture contents (Fig. 4.20). Similar results have been found for other insect species (e.g. Howe 1958). Likewise, George (1985) examined the simultaneous effects of temperature and salinity on the intrinsic rate of population increase of the copepod *Eurytemora herdmani*, which was found to peak at intermediate values of both variables (Fig. 4.21). George argued that if extrapolation was performed to estimate the limits of temperature and salinity permitting population growth, these were generally consistent with reported occurrences of the species. The evidence that this was so, is not, however particularly convincing.

Many studies have also been performed of broad spatial variation in components of local population growth rates of species in the field, although the relation, if any, between these and intrinsic rates of increase is seldom known (and certainly need not be positive). Thus, for many species of birds clutch size increases towards higher latitudes (Fig. 4.22; e.g. Cody 1966; Royama 1969; Ricklefs 1972; Koenig 1984, 1986; Dunn *et al.* 2000; Jenkins and Hockey 2001; but see Järvinen 1986, 1989; Soler and Soler 1992; Yom-tov *et al.* 1994; Reinhardt 1997)[13]; this pattern is also seen for other groups (e.g. Fleming and Gross 1990), and should not be confused with the tendency for an interspecific relationship between clutch size and latitude (e.g. Kulesza 1990; Böhning-Gaese *et al.* 2000). For some of these species there is evidence for an increase in reproductive success towards higher latitudes (e.g. Ricklefs 1972; Corbacho *et al.* 1997; Jenkins and Hockey 2001), in others for a decrease (Reinhardt 1997),

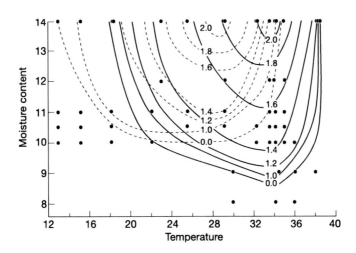

Fig. 4.20 The finite rate of increase (λ) of the beetles *Sitophilus oryzae* (dashed lines) and *Rhizopertha dominica* (solid lines) living in wheat of different moisture contents (per cent) and at different temperatures (°C). From Birch (1953).

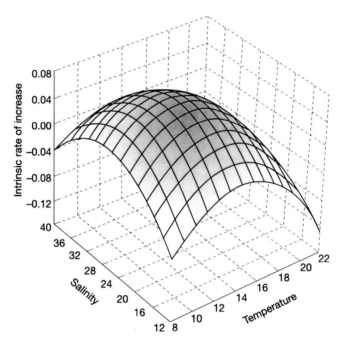

Fig. 4.21   Intrinsic rate of population increase (d⁻¹) of the copepod *Eurytemora herdmani* as a function of temperature (°C) and salinity (⁰/₀₀). From data in George (1985).

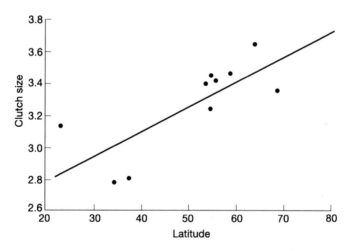

Fig. 4.22   Latitudinal gradient (°N or °S) in clutch size for peregrine falcon *Falco peregrinus*. From Jenkins and Hockey (2001).

and evidence as to whether smaller clutch sizes and lower reproductive rates at lower latitudes are offset by lower mortality rates is mixed (Curio 1989; Bell 1996).

In some cases, there is evidence of a tendency for larger clutches towards the centre of ranges. Thus, Peakall (1970) found that the eastern bluebird *Sialia sialis* in North America has larger clutches at the height of the breeding season in the centre of its range, that the breeding season is longest at mid-latitudes where breeding density is highest, but that there is no geographic pattern in breeding success. Sanz (1997) found that new data from the southern part of the geographic distribution of the pied flycatcher *Ficedula hypoleuca* revealed a quadratic relationship between mean clutch size and latitude (Fig. 4.23), rather than other patterns with latitude that some previous studies had documented. One wonders to what extent this might also be true of some other species for which data have not been obtained over a sufficient range of latitudes, although plainly this is not an issue in many cases. The mean number of fledgling pied flycatchers showed a similar pattern of increase toward

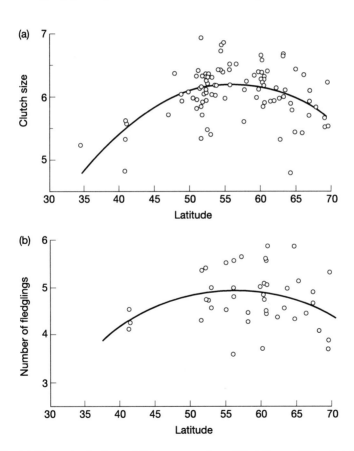

Fig. 4.23   (a) Mean clutch size and (b) mean number of fledglings, of the pied flycatcher *Ficedula hypoleuca* at different latitudes (°N). From Sanz (1997).

mid-latitudes (Fig. 4.23), and Bradford *et al.* (1997) have similarly found that the average abundance of Coho salmon *Oncorhynchus kisutch* smolt produced by streams and rivers in western North America peaked at intermediate latitudes and declined to north and south. Carey *et al.* (1995) found that for the winter annual grass *Vulpia ciliata* ssp. *ambigua* the basic reproductive rate R_o was higher towards the centre of the geographic distribution and also towards the centre of a population.

For many studies of spatial variation in components of population growth rates it is unclear to what extent observed differences simply reflect behavioural responses to environmental conditions and to what extent they are innate. Birch *et al.* (1963) found that at both 20°C and 25°C laboratory populations of *Drosophila serrata* from five localities exhibited innate rates of increase which increased from north to a central peak and then declined again to the south. The conditions under which a species attains its highest rate of increase may change geographically in parallel with environmental conditions. Thus, Bateman (1967) comparing strains of the Queensland fruit fly *Dacus tryoni*, found that at the lowest temperature the Sydney strain had the highest rate of increase and the Cairns strain the lowest, that at intermediate temperatures there were no significant differences, and that at the highest temperature the Cairns strain had the highest rate of increase and the Sydney strain the lowest. Bradshaw *et al.* (2000) examined the performance of nine populations of the pitcher-plant mosquito *Wyeomyia smithii*, from 30° to 50°N when exposed to the same stressful summer temperatures and simulated winter. Survivorship declined with increasing latitude in the summer but not the winter environment, fecundity was not correlated with latitude in either environment, fertility declined with latitude in both, and cohort replacement rate was not correlated with latitude in either environment.

In short, some evidence regarding spatial variation in rates of increase provides support, albeit often tentative, for the vital rates model, but other evidence is contradictory. The model undoubtedly provides a very general caricature of how the abundance structure of ranges is likely to be structured, but how well this caricature matches any given species is unclear. In this sense, like optimum response surface models, its primary value may be heuristic.

Gain and loss surface model

Maurer and Brown (1989) provide a demographic model of the abundance structure of ranges as a mechanistic development of the processes underlying the optimum response surface model of Brown (1984). Local abundance is determined as a function of birth, death, immigration, and emigration rates. Death and emigration rates are treated as increasing functions, birth rates as a unimodal function, and immigration rates as a decreasing function of density. Population gain (births plus immigration) and loss (deaths plus emigration) functions intersect to give a stable equilibrium point, and the distribution of equilibrium points across space generates the abundance structure of the range. If population gain is a non-increasing, or population loss a non-decreasing function of distance from the geographic range centre and of population density, or both, then population density will be a non-increasing function of distance from the range centre (i.e. population density declines towards range edges).

The available evidence for geographic patterns of variation in population processes has already been discussed. Given the difficulty in establishing the existence of even very simple patterns of difference between core and peripheral areas, establishing more complex functions remains largely impossible.

Stochastic temporal variation models

Ives and Klopfer (1997) show that spatial patterns of variation in the local abundance of a species can be generated by stochastic temporal variation in environmental conditions in the absence of any fixed pattern of spatial environmental variation. Their simulation, which bears some features in common with Perry's (1988) models of intraspecific abundance distributions (Section 4.2), comprises populations within cells arranged in a grid, with the dynamics within cells governed by the model

$$x_{t+1} = srf(x_t)x_t + sx_t \qquad (4.18)$$

where x_t is adult density at year t, s is the annual survival of adults, r is the average per capita reproductive rate of adults when population densities are low, and $f(x_t)$ gives the reduction in the reproductive rate as x_t increases. Both r and s are random variables following independent lognormal distributions. In any given year, the random values assigned to r and s are the same within 8×8 blocks of cells, simulating similarity in environmental fluctuations at this scale. However, the locations of what constitutes a block are changed each year, removing any fixed spatial structure. The cells are linked through annual migration, during which 10 per cent of the adults move out of each cell and are redistributed evenly among the eight neighbouring cells. At the edge of the grid individuals can disperse out of the species 'range' and are assumed to die.

This model produced a sigmoidal log abundance–rank abundance plot, covariance in abundances between years separated by moderate lags, strong positive spatial autocorrelation in abundance over moderate distances, and a positive relationship between abundance and distance to the edge of range (the edge of the grid) (Fig. 4.24).

The significance of this model lies in its demonstration that spatial patterns of environmental variation (assumed by most models of abundance surfaces) are not necessary to generate spatial patterns of abundance akin to those that have been observed (although the degree of departure remains debateable; e.g. the spatial auotocorrelations show no marked increase at long lags). However, as Ives and Klopfer (1997) observe, this is not to say that such spatial patterns of environmental variation are not important. Indeed, some combination of the two processes seems likely to be acting.

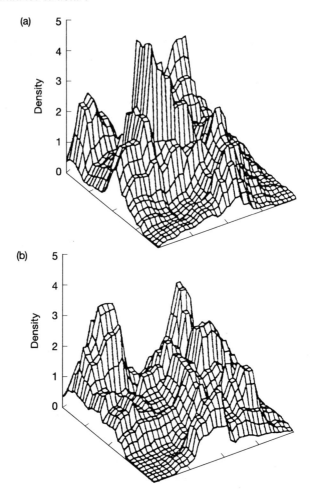

Fig. 4.24 Population densities in each of 625 cells after (a) 180 years, and (b) 200 years of simulation using a stochastic temporal variation model. The horizontal axes represent spatial location in the 25×25 cell grid. From Ives and Klopfer (1997).

4.6 In conclusion

Two principal questions dominate consideration of the abundance structure of the geographic ranges of species. First, what is the general form of this structure? Second, what causes this pattern? Failure to provide a categorical answer to the first question has inevitably severely hampered answers to the second. In particular, concerns have persisted over how best to interpret the surprisingly limited empirical data that are available and thence over attempts to extrapolate more widely. More definitive analyses are plainly required. In the absence of firm conclusions the search for mechanism has tended to assume that the simple scenario of a central peak of abundance with declines towards the range limits pertains more widely.

It is plain that most species are relatively scarce in most of the places in which they occur (with a high proportion of the total number of individuals occurring in a small proportion of the occupied areas), that local abundances show at least a moderate level of positive autocorrelation over short distances and that the level of autocorrelation declines toward moderate ones (what happens beyond this point seems more variable), and that local abundances correlate with spatial variation in some environmental factors. As commonly employed, 'abundance surfaces' are severe abstractions, in that they imply a continuity of variation in local abundance that is seldom met, with all ranges being discontinuous affairs (when examined at even moderate resolutions), in which local populations (for want of a better term) are scattered over the landscape. Multiple hotspots of abundance are frequent, and are often distributed in complex fashion but with local abundances declining away from their immediate vicinity.

These features of the abundance structure of ranges can reasonably be interpreted as local responses to environmental conditions, but with this linkage overlain by the complexities of sink–source dynamics and geographic variation in the influence of these conditions on individuals with different genetic histories. This provides a common connection to mechanistic discussions of the determinants of the limits to species' geographic ranges (Chapter 2) and of the sizes of those ranges (Section 3.6).

A variety of population dynamic models, making different assumptions about dispersal and responses to the environment, produce realistic abundance patterns. Indeed, no simple, overall response to variation in environmental conditions is required to yield patterns of local abundance that are very akin to those commonly observed.

The patterns of abundances of species in space give rise to the abundance structure of local and regional assemblages—this structure is simply the concatenation of the abundances of multiple species (Fig. 4.25). Given that individual species are rare in most places in which they occur, one expects that most of the species that occur

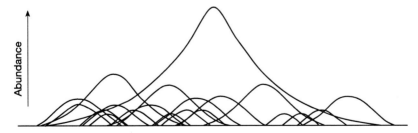

Fig. 4.25 The abundance distribution for an assemblage occurring in any one area results from the positioning of that area relative to the ranges of the different species and the pattern of abundance across those ranges. Smooth hump-shaped abundance functions are used here purely for illustrative purposes.

in any one area will also be at low abundances, and this is indeed the case. Species–abundance distributions (interspecific frequency distributions of abundance) are almost invariably strongly right-skewed, although the statistical models that best describe these patterns are variable and scale-dependent.

In a fascinating, and much neglected paper, Hengeveld *et al.* (1979) explore the relations between the geographic distributions of the abundance of individual species and the abundance structure of the resultant local assemblages. They assume that individuals are independent of one another, that for each species these individuals are spatially distributed according to a defined dispersion function such that density declines monotonically from a central point, and that the centre points for different species are Poisson distributed. Simply by varying the shape of the dispersion function and how this function varies between species, various realistic species–abundance distributions can then be generated, neatly linking the spatial abundance structure of the geographic ranges of individual species and the structure of communities at a variety of spatial scales.

5

Implications

In any general discussion of structure, relating to an isolated part of the universe, we are faced with an initial difficulty in having no a priori criteria as to the amount it is reasonable to expect. We do not, therefore, always know, until we have a great deal of empirical experience, whether a given example of structure is very extraordinary, or a mere trivial expression of something which we may learn to expect all the time.

G. E. Hutchinson (1953)

5.1 Introduction

One of the major, and arguably the most important, driving forces behind recent, on-going, and planned studies in the field of areography is concern about some of the significant environmental problems that humanity is facing at present (Section 1.2.1). These include such issues as the loss of biodiversity, the impacts of climate change, the spread of invasive species, and the control of pests and diseases. Thus far in this volume, despite some occasional remarks, such matters have intentionally been largely ignored. In this concluding chapter, drawing on the material presented in those that preceded it, these issues will be explicitly addressed. Attention is focused particularly on insights related to range contractions and extinctions (Section 5.2), protected areas (Section 5.3), climate change (Section 5.4), aliens (Section 5.5), and reintroductions (Section 5.6). Many of the points that will be made are quite general ones, but are framed here in an applied context.

This exercise highlights a particular case of the observation made by G. E. Hutchinson in the quotation above, in that understanding the appropriate responses to some environmental problems rests fundamentally on identifying the repeatable features of the structure and dynamics of geographic ranges, the variability in those features, and their determinants. As we have seen, in some cases understanding of those features is reasonably advanced. In others it is rather rudimentary.

5.2 Range contractions and extinctions

The 'biodiversity crisis' is a matter of decline and extinction. The populations of many species are declining in numbers of individuals, and local populations are disappearing at high rates, as a result of human activities. In consequence, large numbers of species are regarded as being at high risk of extinction in the near future (Hilton-Taylor 2000), and estimates suggest that species are being lost with a rapidity that is unprecedented outside of the mass extinction events revealed in the fossil record (May *et al.* 1995; Pimm *et al.* 1995). Intraspecific abundance–range size relationships (Section 4.3) provide a useful framework in which to consider the nature of these declines. Thus, Schonewald-Cox and Buechner (1991; see also Wilcove and Terborgh 1984; Lawton 1993; Gaston 1994c) proposed simple, idealized models encompassing four forms that the trajectory to extinction might take, with regard to changes in the size of the geographic range of a species and its total population size, as the total number of individuals declines (Fig. 5.1):

1. *Geographic range size remains approximately constant as the number of individuals declines, and overall density declines with time; there is no density–range size relationship* (Fig. 5.1a). In the limit this model cannot hold, the number of individuals cannot decline very markedly without some loss of occupancy. However, over some span of total numbers of individuals several ways can be envisaged in which these numbers might decline whilst occupancy remains approximately constant, particularly if the sizes of ranges are determined at a coarse spatial resolution (with localized extinctions thus tending to pass undetected), and there is no strong causal link between abundance and range size. Hunting, pollution, and climatic changes, for example, may all potentially result in declines in the numbers of individuals in parts of the range of a species without necessarily leading to local extinctions. However, this seems likely to occur only where the local abundance of a species is initially high or the intensity of these sources of mortality is low, making it probable that the changes may pass unremarked or unnoticed. The fact that species are rare throughout much of their geographic ranges means that in most areas and with even moderate mortality, local extinctions will soon result from the loss of even small numbers of individuals.

2. *Number of individuals and range size decline simultaneously such that density remains roughly constant; there is no density–range size relationship* (Fig. 5.1b). This model fits two likely circumstances in which there is no necessary, causal link between abundance and range size. First, the geographic range of a species may be reduced such that all areas stand a chance of being lost that is directly proportional to their relative extent. This may occur where ranges are eroded by habitat loss, without degradation of the remaining habitat and without differential loss of high or low quality habitat. The draining of individual ponds, for example, may destroy local populations of aquatic species but may neither differentially target populations of a particular size nor have any impact on other local populations (providing these are not linked as a metapopulation, in which the abundance and persistence of different local populations are interdependent; Section 2.3.2). Second, reductions in abundance in high density areas may approximately balance losses (extinction) of lower density areas. Indeed, declines in the local abundances of persistent populations accompanied by the loss of small, and often peripheral, populations appear to be a widespread phenomenon. However, in the

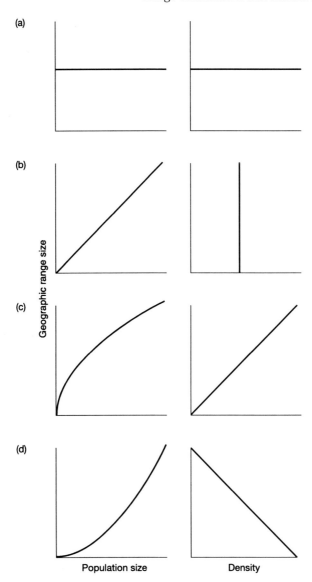

Fig. 5.1 Idealized trajectories to extinction through population size–range size space: (a) geographic range size remains approximately constant as the number of individuals declines, and overall density declines with time; there is no density–range size relationship; (b) number of individuals and range size decline simultaneously such that density remains roughly constant; there is no density–range size relationship; (c) number of individuals and range size decline simultaneously such that the density of individuals declines with time; there is a positive density–range size relationship; and (d) number of individuals and range size decline simultaneously such that the density of individuals increases with time; there is a negative density–range size relationship.

main, it seems that patterns of loss will be unlikely to be so distributed as to maintain average local densities.

3. *Number of individuals and range size decline simultaneously such that the density of individuals declines with time; there is a positive density–range size relationship* (Fig. 5.1c). This seems most likely where the erosion of the geographic range of a species is accompanied by a strong reduction in abundance in areas in which it persists. This could occur in three main ways. First, it might take place in the absence of a causal link between abundance and occupancy, if the forces driving the loss of local populations also result in reductions in abundance in those populations that persist. Broad-scale environmental changes, such as those associated with climate change and atmospheric pollution, might have just this kind of effect. Feedbacks between vegetation and regional climate provide one example of how this could occur. A number of simulation studies using global change models (GCMs) have shown how the loss of areas of vegetation, particularly the loss of forest cover, can change regional climate, because the vegetation itself directly and indirectly influences that climate (through evapotranspiration, albedo effects, etc.), which in turn can change the nature of the remaining vegetation and hence environments for its occupants. Thus, models suggest that the conversion of forest to cropland in the eastern and central United States cooled the climate, resulting in a decrease in mean annual surface air temperature of 0.6 to 1.0°C east of 100°W, with greatest cooling in the Midwest in summer and autumn, when daily mean temperature decreased by $>0.5°$ and 2.5°C, respectively (Bonan 1999). Likewise, the coupling of an atmospheric model and a simple land-surface scheme has indicated that coastal deforestation in West Africa has been a significant contributor to the observed drought in the region (Zheng and Eltahir 1998). In the extreme, Nobre *et al.* (1991) found that when, in a model, the entire Amazonian tropical forests were replaced by degraded grass (pasture), there was a significant increase in the region in mean temperature (about 2.5°C), and a decrease in annual evapotranspiration (30 per cent), precipitation (25 per cent), and runoff (20 per cent). This resulted, in major part, because some 50 to 80 per cent of the precipitation over the Amazonian rain forest is recycled from evapotranspired water (Hayden 1998).

A similar effect might also occur at a smaller scale, because of the influence of habitat fragmentation on conditions within remnant patches that follows from the typical increase in the proportion of 'edge' as patch area declines (e.g. Lovejoy *et al.* 1986; Saunders *et al.* 1991; Sisk and Margules 1993; Kapos *et al.* 1997; Laurance *et al.* 1997; Davies and Margules 1998; Laurance 2000). At such edges, microclimatic changes can be considerable, and can penetrate substantial distances into patches, as evidenced by the demise through desiccation of trees at the margins of tropical forest fragments (Kapos 1989; Laurance *et al.* 1997, 2000). Patterns of predation, parasitism, and invasion can also change dramatically with the creation of greater proportions of edge habitat (e.g. Murcia 1995; Laurance and Bierregaard 1997; McGeoch and Gaston 2000).

Second, in the absence or presence of a causal link between abundance and occupancy, a trajectory of decline down a positive density–range size relationship might occur if areas of higher density have a disproportionate likelihood of loss. This could take place if, for example, these areas undergo differential exploitation, and this exploitation is sufficiently efficient as to result in local extinctions (e.g. complete harvesting of plant populations for horticultural or medicinal purposes). It might also take place if areas in which a species occurs at high abundance are also valuable for other reasons. Indeed, conflict between the persistence of source populations and extractive use is probably common. It is well illustrated by an example from the coastal *Eucalyptus* forests of eastern Australia, described in a series of publications by Braithwaite and colleagues (Braithwaite 1983; Braithwaite *et al.* 1983, 1984, 1988), although it remains to be demonstrated that this does indeed represent a true sink–source system. They found that densities of arboreal mammals, most commonly the greater glider *Petauroides volans*, feathertail glider *Acrobates pygmaeus*, and the sugar glider *Petaurus breviceps*, were highest in certain forest types (tree communities). These types characteristically occur on a particular granitic substratum known locally as Devonian intrusives. Foliage from the trees of these forest types contains higher concentrations of nutrients such as nitrogen, phosphorus, and potassium. By establishing the link between high animal densities, foliage nutrient levels, and geological substratum, they were able to estimate that 63 per cent of the animals occupied a mere 9 per cent of the forest. Remaining animals were spread, still unevenly but in lower densities, throughout another 41 per cent of the forest. About 50 per cent contained no arboreal mammals at all. The source areas for arboreal mammals contain the best timber trees; species such as *Eucalyptus fastigata*, *E. radiata*, *E. cypellocarpa*, and *E. elata*. Such species were recognized as indicators of higher soil fertility by early European settlers and were cleared for farming from all of the broader valleys on which they occurred. The smaller pockets that now remain are of great value to the timber industry and to the arboreal marsupials.

Third, a trajectory of decline down a positive density–range size relationship might occur if there is a causal link between abundance and occupancy, such that the loss of some populations entails a reduction in abundance in those that remain. Such a pattern is likely to result from a very general set of colonization/extinction processes, in which populations can either go extinct from a patch or colonize new patches, and where either of such processes, or both, are related to the local density within patches that are occupied (Fig. 5.2; Warren and Gaston 1997). This does not necessitate that species exhibit a metapopulation structure, but such a structure would tend to meet this set of requirements.

The prevalence of positive intraspecific relationships between local abundance and range size (Section 4.3) suggests that these three forms of dynamics may embrace the most common route toward extinction. Indeed, temporal declines in local abundance and regional occupancy have frequently been observed to be positively correlated (Gaston *et al.* 1998d). Thus, amongst farmland birds in Britain,

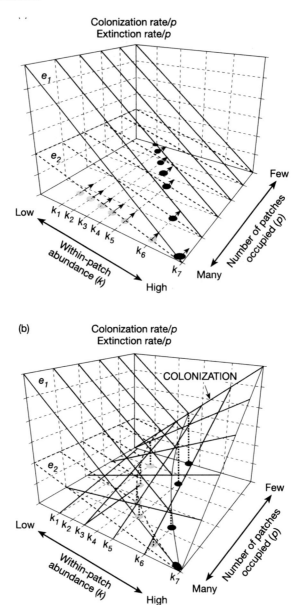

Fig. 5.2 Simple graphical representation of two models of the abundance–occupancy relationship. The basal plane of the graph is the abundance–occupancy relationship, the vertical axis the per-patch extinction and colonization rates. (a) Systems with no external or between-patch colonization. The two planes represent two different extinction rates (e_1 and e_2), where extinction is inversely related to population size (k). Species with characteristic abundances $k_1 \ldots k_7$ go extinct, and hence occupancy decreases per unit time, at different rates, generating a positive abundance–occupancy relationship. Variation among species in the extinction

which have been experiencing dramatic declines in overall population size (some previously common species have declined by more than 50 per cent), these have often comprised both loss of local populations and reductions in density in those that persist (Fuller *et al.* 1995; Siriwardena *et al.* 1998; Gates and Donald 2000). Such dynamics may, nonetheless, be less frequently documented than might otherwise be expected, because lags in the decline of occupancy (that result from declines in density) may make positive intraspecific abundance–occupancy relationships more difficult to detect. Conrad *et al.* (2001) have documented the first clear example of such a lag, for the garden tiger moth *Arctia caja* in the United Kingdom. This species has severely declined throughout the region during the past 35 years, perhaps as a result of poor overwintering survival. Whilst abundance and occupancy decreased greatly between 1968 and 1998, there was a rapid step-change in abundance in 1983 to 1984, followed by a step-change in occupancy in 1987 to 1988.

The likelihood that declines in local density and occupancy will go hand in hand has implications for conservation. These include that local declines in the abundance of rare species (perhaps detected in monitoring schemes) should warn that these may be accompanied by wider changes, and that declining species face a double jeopardy of a progressive lowering density and narrowing occupancy (Lawton 1993).

4. *Number of individuals and range size decline simultaneously such that the density of individuals increases with time; there is a negative density–range size relationship* (Fig. 5.1d). This final trajectory through abundance–range size space will result when areas of habitat in which a species is present, but in which it fails to achieve higher densities, are differentially lost, and there is limited or no impact on abundances elsewhere. Such a pattern may be most likely where the geographic range of a species has one or a few core populations, and other peripheral populations that are maintained by a flow of individuals from these cores (i.e. the peripheral populations are sinks and the core ones are sources). Under this situation the most vulnerable populations are the peripheral ones, but their loss

---

Fig. 5.2 (*continued*) rate–abundance relationship generates variation in this relationship (black circles = extinction rate e_1; grey circles = e_2). (b) Systems where colonization is a function of both local abundance and of the proportion of patches that are empty and extinction is independent of occupancy, but inversely related to abundance. The intersection of the planes indicates the equilibrium occupancy for any given value of k. The resulting abundance–occupancy relationship is indicated on the basal plane of the graph for the seven species with characteristic abundances $k_1 \ldots k_7$ (note that k_1 has no positive equilibrium occupancy in this system). The two extinction–abundance relationships (e_1 and e_2) are as before; black points are the equilibria for e_1, grey points for e_2. It is clear that species-specific variation in the extinction–abundance relationship or the colonization–abundance relationship will generate scatter about the general positive trend, and that more 'extinction resistant' species (having an extinction–abundance relationship like e_2) could occur in the low abundance, high occupancy region of the graph. Even allowing considerable variations in the rates, and the form of the rates (extinction–abundance relationships are unlikely to be linear), a positive relationship of some form is still the expectation. From Warren and Gaston (1997).

has little or no effect on the core populations. The dramatic decline of Kirtland's warbler *Dendroica kirtlandii* may represent an example of just such dynamics. This species is one of the rarest warblers in North America, now breeding solely in north and central Michigan. The population declined from about 500 singing males in 1961 to about 200 in 1971, under pressure from nest parasitism by the brown-headed cowbird *Molothrus ater* (Mayfield 1983). 'When the population crashed in the 1960s, it did not decline evenly across the range but collapsed back into the heart of the range, where the birds nested as densely as ever' (Mayfield 1983, p.975). In 1961, the 500 pairs were distributed over about 16 000 km², and in 1982, 207 pairs were distributed over about 3400 km², giving densities, respectively, of 0.031 and 0.061 pairs/km². Since 1972, a cowbird trapping programme has reduced parasitism from 70 per cent to 3 per cent, and numbers have recovered to 903 singing males in 1999 over an area of 11 700 km² (BirdLife International 2000). This gives an even higher density of 0.077 pairs/km², although the comparability of the area estimate with previous ones is not clear.

The redstart *Phoenicurus phoenicurus* may have passed through a phase of decline in woodland sites in Britain, with similar dynamics. It underwent a decline in occupancy of woodland sites from 1968 to 1978, and subsequently appears to have fluctuated in occupancy without showing any obvious trend (Fig. 5.3; Gaston *et al.* 1999a). In contrast, densities have tended to increase, albeit with substantial variance about the trend. This has led to a negative abundance–occupancy relationship, with no obvious tendency for densities to increase on sites that have remained occupied, and the likelihood that the species has become extinct at sites at which it previously occurred at low densities.

In principle, successful conservation of the density 'hot spots' of the geographic ranges of species in the face of the loss of other areas should result in an increase in mean density with declining occupancy. However, the high densities will only be ephemeral if they were dynamically dependent on other populations that have gone extinct. This emphasizes the need for a regional rather than a site-by-site approach to conservation planning and action, albeit this is at odds with the methodology embodied in some international agreements (e.g. Ramsar Convention on Wetlands of International Importance; Ramsar Convention Bureau 2000) and espoused by some conservation agencies.

5.2.1 Niche and contagion models

In practice, declines to extinction are often likely to walk much more varied paths through abundance–range size space than these simple models might imply, given the complexities of the abundance structure of species geographic ranges and of the processes causing reductions in overall population size. The relative frequency of the different kinds of trajectories, nonetheless, depends, fundamentally, on the ways in which the geographic ranges of species undergo contraction. Attention has focused on two models of this process (Fig. 5.4). The first is closely tied to the idea of a geographic range as an optimum response surface, with more favourable environmental

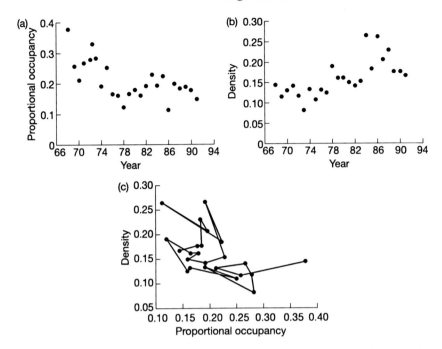

Fig. 5.3 Temporal variation in (a) the proportion of the surveyed sites on which the redstart *Phoenicurus phoenicurus* was found to occur in a given year (1968–1991), (b) the mean local density across woodland sites on which it was recorded to occur in that year, and (c) the relationship between density and occupancy. From Gaston *et al.* (1999a).

conditions, and hence higher local densities, towards the centre of the range and a progressive worsening of these conditions and a resultant decline in local densities towards the range limits (Section 4.5.2). There has been a general expectation that under such circumstances species would disappear first from the periphery of their geographic ranges, and would persist in the core (probably at progressively declining levels of abundance) until the closing stages of range contraction, although dynamic linkages between these populations are generally ignored. However, for a number of endangered species, Channell and Lomolino (2000a, 2000b; Lomolino and Channell 1995, 1998) have found that this pattern of range contraction has not been the case, and that most species have persisted in the periphery of their historical geographic ranges often long after core populations have been driven extinct (Fig. 5.5; see also Fig. 3.1). They interpret this outcome as reflecting a contagion rather than a niche model of decline, in which recent range contraction has been strongly influenced by anthropogenic extinction forces, which mean that those populations that persist longest are simply those that are the last to be affected. Because human activities have tended to sweep directionally across geographic land masses— as exemplified by the spread of croplands and railways across North America,

(a)

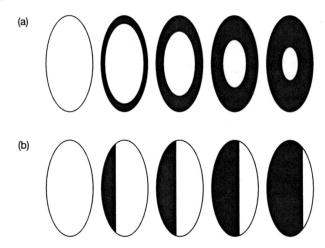

(b)

Fig. 5.4 Idealized examples of (a) niche and (b) contagion models of geographic range contraction. Ranges contract in size from left to right, with the white area indicating the remaining distribution and the black area that from which the species has been extirpated.

although people were, of course, widespread in the region for long before this, (Garrett 1988; Ramankutty and Foley 1999) and the spread of early farming across Europe (Ammerman and Cavalli-Sforza 1971)—peripheral populations most distantly removed from the initial foci of these forces are often the last remnants of species distributions. For example grey wolf *Canis lupus*, mountain lion *Puma concolor*, bison *Bison bison*, elk *Cervus elaphus*, and caribou *Rangifer tarandus*, amongst others, were eliminated from all or most of the eastern United States (Wilcove 2000). The range contractions of large-bodied mammals in the late Quaternary as humans spread across the continents often appear suggestive of such a directional pattern (Flannery 2001; Martin 2001), although the evidence is hotly debated (Grayson 2001).

If a contagion model is indeed the correct one, this would carry some cautionary messages for conservation. Perhaps foremost, it would suggest that the conditions under which remnant populations are found may not be typical of those across which a species was previously distributed and may indeed be very atypical. As previously observed, habitats occupied at range edges may not be typical of those occupied elsewhere in the range, and may indeed not be exploited elsewhere in the range (see Section 2.2.3). In consequence, directing management action at those habitats and maintaining them in the same state may not be the most effective course of action, and translocating individuals to areas with similar habitat may miss valuable opportunities to more effectively increase populations and spread the risk of low numbers (Gray and Craig 1991).

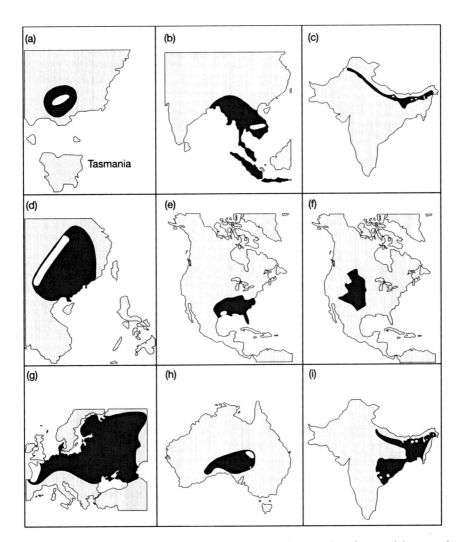

Fig. 5.5 The pattern and range collapse of nine non-volant species of terrestrial mammal (a) Leadbeater's possum *Gymnobelideus leadbeateri*, (b) Javan rhinoceros *Rhinoceros sondaicus*, (c) Indian rhinoceros *Rhinoceros unicornis*, (d) giant panda *Ailuropoda melanoleuca*, (e) red wolf *Canis rufus*, (f) black-footed ferret *Mustela nigripes*, (g) European mink *Mustela lutreola*, (h) dusky hopping mouse *Notomys fuscus*, and (i) water buffalo *Bubalus bubalis*. Historic range in black, extant range in white. From Lomolino and Channell (1995).

5.3 Protected areas

A major instrument for preventing population decline and loss *in situ* is the establishment of a network of protected areas for conservation (nature reserves etc.). Indeed, the creation of such networks is one of the obligations placed on parties to the Convention on Biological Diversity, and to a number of other international treaties and agreements. These areas will of themselves be insufficient, in that they are unlikely ever to cover a large enough extent to embrace viable populations of all the species of concern, and will inevitably be too isolated from one another to enable adequate population rescue effects and other dynamic linkages. However, they provide an important core to a conservation strategy, a set of refugia in which conservation should be the primary concern, and around which other measures can be built (e.g. sympathetic management of areas in the intervening matrix).

Formal analytical approaches to the selection of priority areas for conserving biodiversity provide useful tools in support of this process (Pressey *et al.* 1993; Margules and Pressey 2000), and are finding increasing practical application in exploring the consequences of alternative landuse strategies, particularly as they become better suited to addressing realistic scenarios (e.g. Faith 2000a, 2000b; Rodrigues *et al.* 2000a; Chown *et al.* 2001). However, most of these methods solely employ information on the occurrence of species in different areas (perhaps deriving from the existence or creation of a suitable distribution atlas). They seek to maximize the number of species that can be conserved in the smallest number of areas or overall area, recognizing that funds for conservation are limited and protected areas have to compete with other forms of land use. They also tend to emphasize networks that include a minimum of one representation (i.e. occurrence in at least one, perhaps large, area) of each species.

These features result in some significant limitations, which draw attention to issues of more general import in the structure of protected area networks. First, because species are relatively rare throughout the majority of the area of occupancy of their geographic ranges (Section 4.2), areas selected on the basis of occurrence are on a probabilistic basis alone likely to be those in which a species is not at high abundance. Given that, all else being equal, low abundances are likely to result in a low probability of persistence, this is the very converse of what is desirable. Conservation efforts need to be focused on areas of high abundance, where viability is likely to be good. Of course, for most species and much of the world, data on occurrence are, at best, all that are available as a basis for making conservation decisions. Whilst a hump-shaped model of the abundance structure of species geographic ranges is undoubtedly an extreme caricature of reality (Section 4.5), ensuring that populations well away from range limits are included in networks may provide a rough and ready approach to increasing the likelihood that areas of high abundance are included in networks. But, the complexity of the abundance surfaces of species suggests that any such rule of thumb is likely to be quite poor. There is, however, some intriguing evidence that choosing areas based on the occurrence of hotspots of the abundance of a small number of common and widespread, and more readily censused, species may improve the likelihood that peaks of abundance of other species are also included in a network (Gaston and Rodrigues 2003).

Second, by setting the same minimum representation goal for all species (e.g. the occurrence of a population of each species in at least one of the areas selected), the networks of priority areas that are identified by formal selection techniques will tend to incorporate more representations and more individuals of widely distributed species that are at less threat of extinction (Gaston *et al.* 2000). This occurs because of the existence of the positive relationship between the local densities and the range sizes of species (Section 3.5.9). A single representation of a common species will, on average, include more individuals than will a single representation of a rare species, and this bias is amplified, because the representation targets for common species are more likely to be exceeded (because they will occur at sites chosen for the representation of other species) than will be the case for rare species. These tendencies need to be opposed by setting representation targets that better reflect the conservation importance of the rarer species (e.g. Rodrigues *et al.* 2000b). Such species need often to be protected at multiple sites, not simply for reasons of spreading the risk of extinction, but to ensure that sufficient individuals are secure.

Just how large the global network of protected areas needs to be is a question that has attracted surprisingly little attention, perhaps because the realities of what seems likely to be achieved are expected to fall so far short of any realistic answer, and perhaps because of the difficulty in knowing how to approach the determination of an answer. IUCN-The World Conservation Union advocates that at least 10 per cent of the land area of each nation be set aside for this purpose (IUCN 1993). But although achieving this target would require nearly doubling the currently protected land area (Hobbs and Lleras 1995), concerns have been raised that even this is woefully insufficient and dictated more by considerations of feasibility and politics than of biology (Soulé and Sanjayan 1998). Rodrigues and Gaston (2001) observed that the minimum percentage of area needed to represent all species within a region increases with the number of targeted species, the size of the selection units, and the level of species' endemism. They conclude that, in consequence, (i) no universal target for the size of a network is appropriate, as ecosystems or nations with higher diversity and/or higher levels of endemism require substantially larger fractions of their areas to be protected, (ii) a minimum conservation network sufficient to capture the diversity of vertebrates is not expected to be effective for biodiversity in general, particularly because other major groups are known to have higher levels of endemism, and (iii) the 10 per cent target proposed by IUCN is likely to be wholly insufficient. A meta-analysis of previous studies of the minimum or near-minimum percentage of area required to represent each species in a region at least once, revealed that for selection units of $1° \times 1°$ (*c.* 12 000 km²), the finest resolution that has been considered practical for mapping bird species across an entire continent spanning the tropical zone, 74.3 per cent of the global land area and 92.7 per cent of the tropical rain forests would be required to represent every plant species and 7.7 per cent and 17.8 per cent for higher vertebrates. The values for vertebrates are probably underestimated, given that reserves even of this size ($1° \times 1°$) may not be sufficiently large for maintaining populations of many species (e.g. Newmark 1987, 1996; Mattson and Reid 1991; Nicholls *et al.* 1996).

The third problematic feature of many formal area selection techniques is that maximizing the number of species in the minimum number of areas or the minimum overall area tends to be biased towards selecting as priorities for conservation those areas that lie in ecotones and zones of high habitat heterogeneity (Gaston *et al.* 2000). These are areas in which species are typically quite rare and may often be at the peripheries of their geographic ranges. Thus, in terms of maintaining viable populations of species they are typically not regions of high priority, although they may be important centres of diversification and evolutionary novelty (see Smith *et al.* 1997; Schneider *et al.* 1999). Again ways should be sought to avoid strong biases of this kind in area selection. The particular problem does, however, focus attention on the more general issue of the conservation importance of peripheral populations. Much effort and resources are spent in protecting these populations, particularly for more charismatic groups of species and in more wealthy countries that have low levels of species richness or low levels of endemism. But arguments over the value of peripheral populations are divergent (e.g. Powers 1979; Hunter and Hutchinson 1994; Lesica and Allendorf 1995; Abbitt *et al.* 2000; Rodrigues and Gaston 2002). On the one hand, peripheral populations can be argued to be relatively unimportant, given that they have a high probability of extinction, exhibit low genetic variation, and may divert attention from more significant objectives. On the other hand, these populations may exhibit high levels of local adaptation (which may be valuable in the face of environmental change), be genetically divergent, play significant roles in the evolutionary process (e.g. speciation), and be of conservation importance to local human populations no matter how abundant and widespread the species may be elsewhere. Lesica and Allendorf (1995) argue that it is the degree of genetic drift and the intensity of selection that determine the amount of genetic divergence and should influence the conservation value of peripheral populations (Fig. 5.6). But, much of the conservation attention paid to peripheral populations results from the fundamental mismatch between the patterns of geopolitical boundaries, which strongly influence such decisions, and patterns in the boundaries to species geographic ranges. Species are frequently rare in politically defined spatial units, because the limits to their ranges fall within those units, and they may be given high conservation priority because of this rarity. Schemes have been developed to prioritize the conservation importance of species based on weighting their abundance and occupancy in geopolitical units in relation to their international importance and threat (e.g. Avery *et al.* 1994). This seems the most realistic approach to employ, given resource constraints on conservation.

5.4 Climate change

One of the major challenges for a network of protected areas for the maintenance of biodiversity is how this can be established and developed in such a way that it can

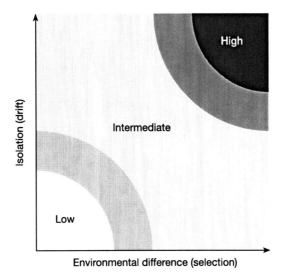

Fig. 5.6 Relative conservation value of peripheral populations from an evolutionary perspective. From Lesica and Allendorf (1995).

accommodate the changes in species distributions that will follow from climate change. The global average surface temperature has increased by $0.6 \pm 0.2°C$ since the late nineteenth century, with warming between 1976 to 2000 being almost global and with the largest increases having occurred over the mid- and high latitudes of the continents in the northern hemisphere (IPCC 2001). It is likely that the rate and duration of the warming in the twentieth century has been larger than at any other time during the last 1000 years, and it has been accompanied by increased annual land precipitation in mid- and high latitudes of the northern hemisphere (with increases also in heavy and extreme precipitation events), decreasing snow cover and extent of land-ice, decreasing amounts of sea-ice in the northern hemisphere, and rises in global mean sea level (IPCC 2001). Such changes will continue, particularly in the absence of any major, concerted actions to mitigate them.

The majority of work on the implications of climate change for species geographic ranges (and thence the ramifications for food security, pest and disease control, conservation, etc.) has comprised modelling present distributions in terms of prevailing environmental conditions, and then examining the consequences of various climate change scenarios based on the modelled links between distribution and climate (Fig. 5.7; e.g. Beerling 1993; Huntley 1994; Brereton *et al.* 1995; Huntley *et al.* 1995; Rutherford *et al.* 1995; Baker *et al.* 1996; Carey 1996; Jeffree and Jeffree 1996; Keleher and Rahel 1996; Nakano *et al.* 1996; Erasmus *et al.* 2000; Rogers and Randolph 2000). Such an approach typically makes a number of critical assumptions, some of which seem unlikely to be met on the basis of present understanding of the role of climate in determining the limits to geographic ranges

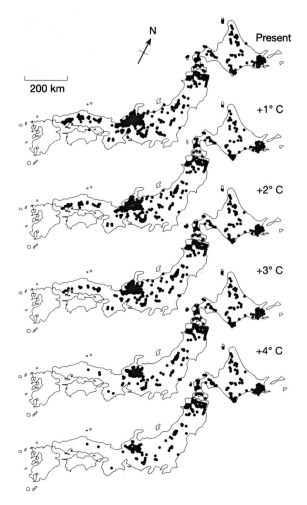

Fig. 5.7 Present and predicted future distributions (for each degree of a 4°C mean annual air temperature increase) of a charr species, Dolly Varden *Salvelinus malma*, in the Japanese archipelago. From Nakano *et al.* (1996).

(Section 2.2.2). These include:

1. *Correlations between climate and the occurrence of a species reflect causal relationships.* As discussed at length earlier (Section 2.2.2), there is no necessity that such correlations be causal. Indeed, such correlations may well reflect relationships between both climate and occurrence and one or more other variables. This need not necessarily matter in the prediction of how a species geographic range will respond to climate change, but only if the assumption is made that all of these relationships remain sufficiently constant as climate alters.

2. *Any influence of other factors on observed relationships between climate and the occurrence of a species, such as resources, competitors, predators, and parasites, will remain constant.* This requires belief in a remarkably simple structure to ecological systems, in which all of these interacting factors respond synchronously to temporal changes in climate in just the same fashion as they vary spatially at any one time. Of course, as the best part of a century of study of ecological systems has shown, synchronous, repeatable responses of this kind do not exist. The microcosm experiments of Davis *et al.* (1998a, 1998b) have demonstrated this in a simple laboratory environment in which most sources of variation have been controlled. Human activities impose a marked influence on the distributions of species, and how these alter with changes in climate is alone likely to be extremely complicated, and dependent on social pressures and technological developments.

3. *Temporally generalized climatic conditions (seasonal or annual means, medians, etc.) are more important influences on the distributions of species than rates of climatic change and extreme events.* As has previously been observed, rates of change and extreme events may actually be quite significant determinants of species distributions (Section 2.2.2), particularly because they influence the effectiveness of behavioural and physiological responses.

4. *Spatially generalized climatic conditions (based on conditions at the closest climate stations, or spatial interpolations of conditions at these stations) sufficiently characterize the conditions that individuals of a species actually experience.* Given the enormous heterogeneity of microclimates, and the abilities of organisms to avoid extremes and be buffered from variation by occupying particular environments, such generalized measures can, at best, be crude caricatures or correlates of the conditions actually experienced, and of the likely impacts of changes in these conditions.

5. *Climate change will be relatively simple, inasmuch as its influence on species distributions can be summarized in terms of the projected changes in one or a few variables.* In fact, several components of climate are expected to change, and the responses of organisms may be sensitive to interactions between these components; there need be no exact analogues of present climates in the future (Lawton 1995).

6. *There is no physiological capacity to withstand environmental conditions which are not components of those prevailing within the areas in which a species presently occurs.* In many of the studies of physiological tolerance and capacity that have been conducted, this has emphatically been found not to be the case (Spicer and Gaston 1999). Arguably, rather than the fit of species to their environments, what is more striking is the widespread ability to withstand conditions that apparently lie well beyond those that they are ever likely to encounter. Much of the confusion on this point has arisen through undue emphasis on the average physiological traits of individuals in one or more local populations rather

than the frequency distributions of those traits (Spicer and Gaston 1999). There is often substantial variation in, say, the physiological tolerances of individuals within a population, with those of the more extreme individuals being substantially greater than the mean.

7. *Range shifts, expansions, or contractions are not accompanied by physiological changes, other than local non-genetic acclimatization, and thus present responses to climate will provide an accurate indication of future responses.* It would seem likely that faced with a novel set of environmental conditions, local genetic adaptation could occur. After all, there is evidence that peripheral populations can exhibit such local adaptation (Chapter 2). However, it remains an open question as to the relative rates at which climate will change and evolutionary responses might occur. Long-term experiments are required, examining the physiological responses of populations over multiple generations to slow (by laboratory experimental standards, if not by global ones) directed change in, say, temperature. Whilst they may still not be realistic, conducting such experiments using rates of temperature change that are an order of magnitude slower than have been employed in the past is possible.

An example of an adaptive evolutionary response to global warming comes from the field. Over the last 30 years, the genetically controlled photoperiodic response of the pitcher-plant mosquito *Wyeomyia smittis* has shifted toward shorter, more southern daylengths as growing seasons have become longer (Bradshaw and Holzapfel 2001). This response has been faster in northern populations where selection is stronger and genetic variation greater.

8. *Dispersal limitation is unimportant both in the determination of the present distributions of species and in their ability to respond to changes in climate.* The first of these points presumes that ranges are in equilibrium with climatic conditions, the second that they will remain so. The frequency with which geographic ranges are presently dispersal limited is difficult to evaluate, as this is a question of degree rather than absolutes (Section 2.2.1). All species are dispersal limited in as much as there is probably a region of the world in which environmental conditions are suitable for establishment but which they have yet to reach. To be fair, the issue of how quickly the ranges of species can respond to climate change has been much debated. Given that human-engendered climate change is forecast to occur at rates greater than those seen in many historic periods, this suggests the potential for mismatches between distributions and climate to become more common and more exaggerated. The tracking of projected rates of climate change would require migration rates that may be an order of magnitude faster than those seen at the end of the Pleistocene glaciations (e.g. Huntley 1991, 1994). Moreover, the situation is made immeasurably worse by the dramatic habitat changes that human activities have wrought, resulting in the fragmentation of many vegetation types, and the isolation of patches often by substantial distances from their nearest neighbours.

In short, a number of critical assumptions of climate matching approaches to predicting the response of species geographic ranges to climate change are likely to be severely violated.

Of course, none of this is to say that the geographic ranges of species will not respond to change in climatic conditions. They already are responding, and will continue to do so, just as they have throughout the history of life on Earth. Rather, the changes that will occur may be substantially more difficult to predict than simple climate matching approaches might suggest. Studies of the determinants of the limits to the geographic ranges of species have taught us that unravelling mechanisms is often very difficult, that these limits are often set by complex interactions between multiple factors, that these factors may change radically from one part of the range periphery to another, and that even subtle changes in conditions may have dramatic influences on occupancy. The responses of even closely related, and physiologically and ecologically very similar, species may thus be wildly different and difficult to predict. They will move in different directions, at different rates, and to different extents, and may from many perspectives appear to behave idiosyncratically. Certainly, simple models based on climate matching approaches are likely to prove misleading.

Unfortunately, it is not obvious what should replace simple climate matching models. A parallel can, perhaps, be drawn with population dynamics models. In order to predict successfully the population dynamics of a single species, particularly where the consequences of different management options need to be evaluated (some of which may have significant financial implications), detailed models are often constructed embodying the best understanding of crucial features of the species biology. These models may be data demanding but have, on occasion, proven very valuable tools. However, they are specific, not readily modified to predict the dynamics of the populations of other species or even the same species in other areas, and are not the most effective route to understanding the broad principles of population dynamics. The general effects of, say, different kinds of age structures, of different spatial distributions of subpopulations, or of the influence of competitors and predators are more appropriately examined using, often simpler, more generic models. In its details, the population dynamics of every species may differ, but in terms of broad patterns they can all be divided into just a few classes of dynamics (Lawton 1992).

Climate matching approaches to determining the consequences of climate change for the distributions of species are more akin to the simpler generic population dynamic models. They incorporate very little detailed information on the individual species, and rely for their predictions on some greatly simplifying assumptions. They may, nonetheless, provide some general indications of how species distributions will change. What seems unlikely is that they will provide good predictions of how individual species will actually respond on the ground. For this, more detailed models will be required that incorporate far more information on the biologies of those particular species, and make fewer simplifying assumptions. A particular difficulty at present is that the more generic approach is being used to make quite specific predictions at the level of the individual species.

The consequence of the divergent responses of species to climate change will ultimately be the wholesale reshaping of assemblages. Simple vegetation models suggest that the prevailing biomes over much of the land surface will change with climate (Leemans 1996). But this obscures much important detail. In any given region, some species will become extinct, some that were rare will become more abundant and widespread, some that were abundant will become scarce and restricted, and yet others will colonize from elsewhere. These changes will not take place independently, but will be influenced by the endless shifting of direct and indirect, and positive and negative, species interactions.

5.5 Aliens

An issue of major concern in the context of climate change is the effect this may have on patterns of accidental or intentional introduction of alien species. Already, 400 000 species are estimated to have been accidentally or intentionally introduced by human activities to areas that lie beyond the natural limits to their geographic ranges (Pimentel 2001), and in some regions the dispersal of species associated with the transport of people and goods may now be far more important determinants of their occurrence than any natural dispersal mechanisms (Hodkinson and Thompson 1997). Whilst a high proportion of such introductions have no evident adverse impacts (Williamson 1996), the damage wrought by others can be immense. Pimentel *et al.* (2000) estimate that the approximately 50 000 non-indigenous species in the United States alone result in economic damage and control of $137 billion per annum. Although one can question the details of their calculation, the magnitude of the problem is inescapable.

Paralleling the case for range contractions and extinctions (Section 5.2), intraspecific abundance–range size relationships again provide a useful framework in which to consider the nature of the spread of alien species on introduction to a new region, and some of the associated issues. Indeed, there are counterparts of each of Schonewald-Cox and Buechner's (1991) four idealized models of trajectories to extinction. Unfortunately, few studies have reported data appropriate for testing these models, largely because whilst local abundances may be documented for alien species early in the invasion process, only areas of occupancy are documented later.

1. *Overall range size remains approximately constant as the number of individuals increases, and overall density increases with time; there is no density–range size relationship.* There are two related, but rather limited, circumstances in which this seems most likely to occur. First, this may take place during the lag phase when the population of a newly introduced species is building prior to its wider expansion (Section 3.4.2), and where a previously unproblematic species experiences some ecological release such that its numbers build up prior to its spreading more

widely. Such a dynamic seems unlikely to persist for long without at least some small expansion of range, although if range sizes are measured at a sufficiently crude and low resolution such changes may pass undetected. Second, increase in abundance with little change in range size may occur toward the end of an invasion, when the full extent of distribution has been attained, but local populations continue to grow (Maurer *et al.* 2001).

2. *Number of individuals and range size increase simultaneously such that density remains roughly constant; there is no density–range size relationship.* This could arise either if densities build up in previously occupied areas as new areas are colonized at initially low densities, or densities build up very rapidly in newly colonized areas such that average density does not change markedly. The latter may be facilitated by any tendency for early periods of colonization to be associated with an escape from constraints on population increase, such as predators or parasites, or an over-exploitation of resources (e.g. Chiang 1961; Caughley 1970). However, both mechanisms require a balance between dynamic processes keeping density constant that is unlikely to occur very widely.

3. *Number of individuals and range size increase simultaneously such that the density of individuals increases with time; there is a positive density–range size relationship.* This will occur whenever the expansion of the geographic range size of a species is accompanied by increases in density in those areas in which populations are already established. This seems likely to be the most general pattern for natural colonists. Thus, toward the end of the last century, the fulmar *Fulmarus glacialis* became well-established in Britain and has subsequently massively increased in abundance and distribution. Over the period for which comparable data are available, 1878 to 1960, the number of nests increased at a faster rate than the number of colonies (Newton 1997, and references therein). Similarly, Hengeveld (1989) finds such a pattern in his collation of estimates of the number of individuals of, and number of sites occupied by, the collared dove *Streptopelia decaoto* through time as it invaded Britain. The number of individuals increased more rapidly than the number of locations, implying an increasing density (Fig. 5.8). Whether a similar pattern occurs for species that have been accidentally or intentionally introduced is more difficult to demonstrate using available information; however it seems highly likely, requiring, principally, that range expansion takes place before local populations in previously established areas have attained the limits to their density.

4. *Number of individuals and range size increase simultaneously such that the density of individuals decreases with time; there is a negative density–range size relationship.* This situation can occur where, as the overall population increases, numbers of potential colonists spread out over large areas, generating increasingly low densities. The common grackle *Quiscalus quiscula* (whose distribution is illustrated in Fig. 4.10) may provide an example of such dynamics (Fig. 5.9; Gaston and Curnutt 1998).

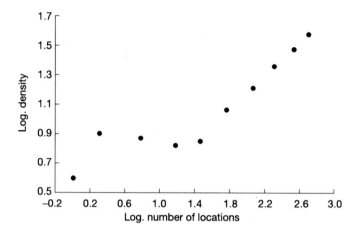

Fig. 5.8 The relationship between the mean abundance of the collared dove *Streptopelia decaocto* in Great Britain and the number of locations at which it was recorded as it spread, between 1955 (smallest number of localities) and 1964 (highest number). From data in Hengeveld (1989).

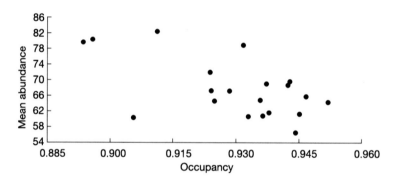

Fig. 5.9 The relationship between the mean abundance of the common grackle *Quiscalus quiscula* across all the survey routes on which it was recorded to occur in a given year (1970–89), and the proportion of all the surveyed routes in the United States and Canada at which it was found to occur that year. From Gaston and Curnutt (1998).

Again, paralleling discussion of range contraction (Section 5.2), one can contrast a niche-based and a contagion model of geographic spread. Under a niche-based model, a species would establish first in those areas with the most favourable environmental conditions, and hence in those in which, given time, peaks of abundance in the geographic range would form. Spread from these core areas would thence be into less favourable environmental conditions. Under a contagion model, a species would likely establish first in what would later become a peripheral area of the range, because this would be the closest area to the native range and thus more likely to receive introduced individuals (either by natural processes or as a result of human activities). Spread from those peripheral areas would thence be into a mixture of what would later become core and peripheral parts of the introduced geographic range of the species.

5.6 Reintroductions

Not all introductions are undesirable. Where the forces that have caused populations severely to decline or to be extirpated in a part of its historical geographic range have been removed or mitigated, the possibility may arise to reintroduce or translocate individuals of a species from elsewhere (Fischer and Lindenmayer 2000). Thus, for example, in Scotland animal species for which reintroductions have taken place or have been discussed include white-tailed eagle *Haliaeetus albicilla* (reintroduced), goshawk *Accipiter gentilis* (reintroduced), capercaillie *Tetrao urogallus* (reintroduced), red kite *Milvus milvus* (reintroduced), European beaver *Castor fiber* (proposed), wolf *Canis lupus* (highly unlikely), European lynx *Lynx lynx* (unlikely), and wild boar *Sus scrofa* (possible).

Some messages regarding such actions arise from consideration of the structure of the geographic ranges of species. First, reintroductions to peripheral areas of geographic ranges are, all else being equal, probably the most likely to fail. Wolf *et al.* (1998) provide empirical evidence on this point, in their analyses of the predictors of success in avian and mammalian translocations. Employing both non-phylogenetic multiple regression and multiple regression using phylogenetically independent contrasts, they found that the release of translocated animals into the core of a species' historical range was a consistent predictor of success.

Second, as a general rule, reintroductions should avoid the use of individuals from peripheral areas of geographic ranges. Given that there are patterns of genetic variation and novelty across the geographic ranges of species, particularly between core and peripheral areas (Section 2.4), then other considerations aside it is logical preferentially to translocate individuals from areas close to or more similar to the target areas. If translocations are preferentially targeted at areas core to species' historical geographic ranges, then this amounts to avoiding translocating individuals from peripheral populations.

5.7 Final words

The biosphere is sandwiched between the lithosphere and the atmosphere, comprising a tiny proportion of the summed depth of all three. This thin layer of life is made up of the distributions of many millions of species. These geographic ranges are not scattered at random, but form a richly textured surface that gives rise to the patterns of biodiversity, with peaks of high species richness, valleys of low richness, and ample plains in between. The range of each species in turn can be viewed as a surface, comprised of variation in the numbers of individuals in different areas, with its own unique patterning of peaks and valleys, sculpted in part by variation in the abundances of other species, serving as resources, competitors, predators, or parasites. Other, less evident, surfaces of variation in genetics and life history exhibit more stable but nonetheless complex behaviours. All of these surfaces ebb and flow by seasons and by years, and have done so since life first established a firm foothold on Earth.

To the terrestrial-bound observer, this broad picture is difficult to visualize. However, the patterning of occurrence, abundance, genetic variation, and so forth across the geographic ranges of large numbers of species gives rise to the structures of the local assemblages that can readily be seen, and with which more direct interaction can take place. The ebbs and flows in these surfaces reflect the births and deaths of individual organisms in the numerous, small patches of habitat that surround our every-day lives, and which we traverse when venturing further afield. Equally, of course, local population establishment, breeding success, and patterns of differential survival all contribute to forming geographic scale variation in biodiversity and ecology. Moreover, processes at the two scales, the local and the regional, feed back on one another. Local assemblage structure depends in part on geographic range structure, which determines, for example, the potential for a local site to be colonized and the genes carried by the individuals that will arrive.

Human activities commonly disrupt local assemblages, driving some species extinct, creating opportunities for others, and changing the abundances of most. But, these effects cannot be viewed in such isolation. These disruptions alter the shapes of the surfaces that are the geographic ranges of species, the consequences often rippling outwards to cause yet further changes in those shapes. If we are to understand fully how humanity is influencing biodiversity we need to understand those ripples. We need to understand the structure of geographic ranges.

Notes

To improve the readability of the main body of the text of this book, limits have been imposed on the numbers of studies cited in support of particular statements. To maintain a rather more complete coverage of the literature, particularly sources of useful historical or empirical examples, additional studies are listed in this section, with their presence being indicated in the text by superscript numbers.

Chapter 1

1. Sharrock 1976; Heath *et al.* 1984; Lack 1986.

Chapter 2

1. Cook 1924; Salisbury 1932; Matthews 1937; Pepper 1938; Andrewartha and Birch 1954; Chiang 1961; Carne 1965; Crisp 1965; MacArthur 1972.

2. Pepper 1938; Iversen 1944; Sealy and Webb 1950; Dahl 1951, 1992; Voous 1960; Carne 1965; Davison 1970; Willis 1985; Grace 1987; Pekkarinen 1989; Taulman and Robbins 1996.

3. Kohlmann *et al.* 1988; Carrascal *et al.* 1993; Rogers and Williams 1994; Beerling *et al.* 1995; Carey *et al.* 1995; Huntley *et al.* 1995; Shao and Halpin 1995; Baker 1996; Romero and Real 1996; Beard *et al.* 1999; Thomas and Lennon 1999; Venier *et al.* 1999; Cumming 2000.

4. Twomey 1936; Pepper 1938; Panetta and Dodd 1987.

5. Brereton *et al.* 1995; Huntley *et al.* 1995; Rutherford *et al.* 1995; Baker *et al.* 1996; Keleher and Rahel 1996.

6. Stephen 1938; Southward and Crisp 1954; Crisp 1965; Pollard 1979; Ford 1982; Tiainen *et al.* 1983; Diamond 1984; Emlen *et al.* 1986; Kozár and Dávid 1986; Pekkarinen 1989; Taulman and Robbins 1996; Lindgren *et al.* 2000.

7. Newton 1998; Spicer and Gaston 1999.

8. Eckert and Barrett 1993; Marren 1999; García *et al.* 2000; Jones *et al.* 2001.

9. Ryrholm 1989; Hengeveld 1990a; Dekker and Beukema 1993.

10. Crowson 1981; Hengeveld 1990a; Burban et al. 1999.

11. Blackburn and Duncan 2001a.

12. Machado-Allison and Craig 1972; Fukuda and Grant 1980; Gilbert 1980; Chapin and Chapin 1981; Grace 1987; Armbruster *et al.* 1998; Spicer and Gaston 1999; Hewitt 2000.

13. Gram and Sork 1999; Lammi et al. 1999; Jenkins and Hoffmann 2000.

Chapter 3

1. Gaston 1994a; Maurer 1994; Brown *et al.* 1996.

2. Hengeveld and Hogeweg 1979; Williams 1988, 1991; Owen and Gilbert 1989; Ford 1990; Buzas and Culver 1991; Lahti *et al.* 1991; Quinn *et al.* 1996; Gaston and Blackburn 2000; Simková *et al.* 2001.

3. Gaston 1994a, 1996c; Brown *et al.* 1996.

4. Tauber and Tauber 1989; Ripley and Beehler 1990; Abbott 1992; Schliewen *et al.* 1994; Lazarus *et al.* 1995; Rosenzweig 1995; Higashi *et al.* 1999.

5. Barton and Charlesworth 1984; Mayr 1988; Taylor and Gotelli 1994; Wagner and Erwin 1995; Barraclough and Vogler 2000.

6. Strong *et al.* 1984; Zwölfer 1987; Straw 1994.

7. Plants (Gotelli and Simberloff 1987; Collins and Glenn 1990, 1997; Rees 1995; Boeken and Shachak 1998; Thompson *et al.* 1998), spiders (Pettersson 1997), bryozoans (Watts *et al.* 1998), grasshoppers (Kemp 1992; Collins and Glenn 1997), scale insects (Kozár 1995), hoverflies (Owen and Gilbert 1989), bumble-bees (Obeso 1992; Durrer and Schmid-Hempel 1995), macromoths (Inkinen 1994; Quinn *et al.* 1997b), butterflies (Hanski *et al.* 1993; Thomas *et al.* 1998; Cowley *et al.* 2001), beetles (Nilsson *et al.* 1994), bracken-feeding insects (Gaston and Lawton 1988), frogs (B.R. Murray *et al.* 1998), birds (Fuller 1982; Hengeveld and Haeck 1982; O'Connor and Shrubb 1986; O'Connor 1987; Gaston and Lawton 1990; Sutherland and Baillie 1993; Gregory 1995; Gaston and Blackburn 1996b; Blackburn *et al.* 1997a, b, 1998b; Collins and Glenn 1997; Gaston *et al.* 1997b, c, 1998e; Newton 1997; Poulsen and Krabbe 1997; Gregory *et al.* 1998; Cornelius *et al.* 2000; Fernández-Juricic 2000; Hughes 2000), and mammals (Brown 1984; Blackburn *et al.* 1997b; Collins and Glenn 1997; Johnson 1998a).

Chapter 4

1. Pielou 1977b; Strayer 1991; Gaston 1994a; Winston and Angermeier 1995; Murray *et al.* 1999; Condit *et al.* 2000; Watkinson *et al.* 2000.

2. Gaston 1994a.

3. Hassell *et al.* 1987; Rosewell *et al.* 1990.

4. Plants (Boecken and Shachak 1998), butterflies (Pollard *et al.* 1995; van Swaay 1995), moths (Conrad *et al.* 2001), fish (Winters and Wheeler 1985; Crecco and Overholtz 1990; MacCall 1990; Rose and Leggett 1991; Swain and Wade 1993; Swain and Sinclair 1994), and birds (Gibbons *et al.* 1993; Smith *et al.* 1993; Ambrose 1994; Tucker and Heath 1994; Fuller *et al.* 1995; Hinsley *et al.* 1996; Cade and Woods 1997; Gaston *et al.* 1997a, 1998d, 1999b; Newton 1997; Blackburn *et al.* 1998b; Díaz *et al.* 1998; Donald and Fuller 1998; Gaston and Curnutt 1998; Venier and Fahrig 1998; Tellería and Santos 1999; Gaston and Blackburn 2000).

5. Blackburn *et al.* 1998b; Gaston and Curnutt 1998.

6. Dekker and Beukema 1993; Lowe and Hauer 1999; Vander Haegen *et al.* 2000; Yagami and Goto 2000; Pearce and Ferrier 2001.

7. Brussard 1984; Hengeveld 1990a; Maurer 1999.

8. Plants (Salisbury 1932; Huntley *et al.* 1989; Carey *et al.* 1995), insects (McClure and Price 1976; Randall 1982; Brussard 1984; Svensson 1992; Whitcomb *et al.* 1994; Nash *et al.* 1995), molluscs (Kiflawi *et al.* 2000), fish (Macpherson 1989), birds (Dow 1969; Bock *et al.* 1977; Emlen *et al.* 1986; Bart and Klosiewski 1989; Tellería and Santos 1993; Maurer 1994; Maurer and Villard 1994; Brown *et al.* 1995; Curnutt *et al.* 1996), mammals (Caughley *et al.* 1988; Kauhala 1995).

9. van Rossum *et al.* 1997; Virgós and Casanovas 1999; Pérez-Tris *et al.* 2000; Jones *et al.* 2001.

10. Salisbury 1932; Brussard 1984; Wiens 1989; Woods and Davis 1989; Nash *et al.* 1995; Quinn 1995; Pérez-Tris *et al.* 2000.

11. Kendeigh 1974; Brussard 1984.

12. Kieckhefer and Elliott 1989; Kieckhefer *et al.* 1989; Michels and Behle 1989; Yaninek *et al.* 1989; Smith 1992.

13. Bahus 1993; Young 1994; Baker 1995; Bell 1996; Corbacho *et al.* 1997; Hendricks 1997.

References

Abbitt, R. J. F., Scott, J. M. and Wilcove, D. S. (2000). The geography of vulnerability: incorporating species geography and human development patterns into conservation planning. *Biological Conservation*, **96**, 169–175.

Abbott, R. J. (1992). Plant invasions, interspecific hybridization, and the evolution of new plant taxa. *Trends in Ecology and Evolution*, **7**, 401–405.

Acreman, S. J. and Dixon, A. F. G. (1989). The effects of temperature and host quality on the rate of increase of the grain aphid (*Sitobion avenae*) on wheat. *Annals of Applied Biology*, **115**, 3–9.

Addo-Bediako, A., Chown, S. L. and Gaston, K. J. (2000). Thermal tolerance, climatic variability and latitude. *Proceedings of the Royal Society, London*, B **267**, 739–745.

Agnew, A. D. Q. (1968). Variation and selection in an isolated series of populations of *Lysimachia volkensii* Engl. *Evolution*, **22**, 228–236.

Ågren, J. (1996). Population size, pollinator limitation, and seed set in the self-incompatible herb *Lythrum salicaria*. *Ecology*, **77**, 1779–1790.

Alkon, P. U. and Saltz, D. (1988). Foraging time and the northern range limits of the Indian crested porcupine (*Hystrix indica* Kerr). *Journal of Biogeography*, **15**, 403–408.

Allen, C. R., Pearlstine, L. G. and Kitchens, W. M. (2001). Modeling viable mammal populations in gap analyses. *Biological Conservation*, **99**, 135–144.

Allsopp, P. G. (1999). How localized are the distributions of Australian scarabs (Coleoptera, Scarabaeoidea)? *Diversity and Distributions*, **5**, 143–149.

Ambrose, S. J. (1994). The Australian bird count: a census of the relative abundance of common land birds in Australia. In *Bird numbers 1992. Distribution, monitoring and ecological aspects*. Proceedings of the 12th International Conference of the IBCC and EOAC, Noordwijkerhout, The Netherlands (eds W. J. M. Hagermeijer and T. J. Verstrael), pp. 595–606. Statistics Netherlands, Voorburg/Heerlen and SOVON, Beek-Ubbergen, The Netherlands.

Ammerman, A. J. and Cavalli-Sforza, L. L. (1971). Measuring the rate of spread of early farming in Europe. *Man*, **6**, 674–688.

Andersen, J. (1993). Beetle remains as indicators of the climate in the Quaternary. *Journal of Biogeography*, **20**, 557–562.

Andersen, J. (1996). Do beetle remains reliably reflect the macroclimate in the past?—a reply to Coope and Lemdahl. *Journal of Biogeography*, **23**, 115–121.

Andersen, M., Thornhill, A. and Koopowitz, H. (1997). Tropical forest disruption and stochastic biodiversity losses. In *Tropical forest remnants: ecology, management, and conservation of fragmented communities* (eds W. F. Laurance and R. O. Bierregaard Jr), pp. 281–291. University of Chicago Press, Chicago.

Anderson, R. M., Gordon, D. M., Crawley, M. J. and Hassell, M. P. (1982). Variability in the abundance of animal and plant species. *Nature*, **296**, 245–248.

Anderson, S. (1977). Geographic ranges of North American terrestrial mammals. *American Museum Novitates*, **2629**, 1–15.

Anderson, S. (1984a). Geographic ranges of North American birds. *American Museum Novitates*, **2785**, 1–17.

Anderson, S. (1984b). Areography of North American fishes, amphibians and reptiles. *American Museum Novitates*, **2802**, 1–16.

Anderson, S. (1985). The theory of range-size (RS) distributions. *American Museum Novitates*, **2833**, 1–20.

Anderson, S. and Marcus, L. F. (1992). Areography of Australian tetrapods. *Australian Journal of Zoology*, **40**, 627–651.

Andrewartha, H. G. and Birch, L. C. (1954). *The distribution and abundance of animals*. University of Chicago Press, Chicago.

Anon (1999). The UK and European data holdings. http://www.nmw.ac.uk/ite/edn2.html.

Anscombe, F. J. (1948). The statistical analysis of insect counts based on the negative binomial distribution. *Biometrics*, **5**, 165–173.

Antonovics, J. (1976). The nature of limits to natural selection. *Annals of the Missouri Botanical Garden*, **63**, 224–247.

Araújo, M. B. and Williams, P. H. (2000). Selecting areas for species persistence using occurrence data. *Biological Conservation*, **96**, 331–345.

Araujo, R., Ramos, M. A. and Molinet, R. (1999). Growth pattern and dynamics of a southern peripheral population of *Pisidium amnicum* (Muller, 1774) (Bivalvia: Sphaeriidae) in Spain. *Malacologia*, **41**, 119–137.

Archibald, J. D. (1993). The importance of phylogenetic analysis for the assessment of species turnover: a case history of Paleocene mammals in North America. *Paleobiology*, **19**, 1–27.

Arita, H. T. (1993). Rarity in Neotropical bats: correlations with phylogeny, diet, and body mass. *Ecological Applications*, **3**, 506–517.

Arita, H. T., Robinson, J. G. and Redford, K. H. (1990). Rarity in Neotropical forest mammals and its ecological correlates. *Conservation Biology*, **4**, 181–192.

Armbruster, P., Bradshaw, W. E. and Holzapfel, C. M. (1998). Effects of postglacial range expansion on allozyme and quantitative genetic variation of the pitcher-plant mosquito, *Wyeomyia smithii*. *Evolution*, **52**, 1697–1704.

Armstrong, J. T. (1965). Breeding home range in the nighthawk and other birds; its evolutionary and ecological significance. *Ecology*, **46**, 619–629.

Arnold, H. R. (1993). *Atlas of mammals in Britain*. HMSO, London.

Arnold, H. R. (1995). *Atlas of amphibians and reptiles in Britain*. HMSO, London.

Arrontes, J. (1993). Nature of the distributional boundary of *Fucus serratus* on the north shore of Spain. *Marine Ecology Progress Series*, **93**, 183–193.

Arroyo, M. T. K., Riveros, M., Peñaloza, A., Cavieres, L. and Faggi, A. M. (1996). Phytogeographic relationships and regional richness patterns of the cool temperate rainforest flora of southern South America. In *High-latitude rainforests and associated ecosystems of the west coast of the Americas: climate, hydrology, ecology, and conservation* (eds R. G. Lawford, P. B. Alaback and E. Fuentes), pp. 134–172. Springer, New York.

Asher, J., Warren, M., Fox, R., Harding, P., Jeffcoate, G. and Jeffcoate, S. (2001). *The millennium atlas of butterflies in Britain and Ireland*. Oxford University Press, Oxford.

Atkinson, R. P. D., Briffa, K. R. and Coope, G. R. (1987). Seasonal temperatures in Britain during the past 22,000 years, reconstructed using beetle remains. *Nature*, **325**, 587–592.

Austin, M. P. (1976). On non-linear species response models in ordination. *Vegetatio*, **33**, 33–41.

Austin, M. P. (1987). Models for the analysis of species' responses to environmental gradients. *Vegetatio*, **69**, 35–45.

Austin, M. P., Cunningham, R. B. and Fleming, P. M. (1984). New approaches to direct gradient analysis using environmental scalars and statistical curve-fitting procedures. *Vegetatio*, **55**, 11–27.

Austin, M. P. and Gaywood, M. J. (1994). Current problems of environmental gradients and species response curves in relation to continuum theory. *Journal of Vegetation Science*, **5**, 473–482.

Austin, M. P., Groves, R. H., Fresco, L. M. F. and Kaye, P. E. (1985). Relative growth of six thistle species along a nutrient gradient with multispecies competition. *Journal of Ecology*, **73**, 667–684.

Austin, M. P., Nicholls, A. O., Doherty, M. D. and Meyers, J. A. (1994). Determining species response functions to an environmental gradient by means of a β–function. *Journal of Vegetation Science*, **5**, 215–228.

Avery, M., Gibbons, D. W., Porter, R., Tew, T., Tucker, G. and Williams, G. (1994). Revising the British Red Data List for birds: the biological basis for U. K. conservation priorities. *Ibis*, **137**, S232–S239.

Avnir, D., Biham, O., Lidar, D. and Malcai, O. (1998). Is the geometry of nature fractal? *Science*, **279**, 39–40.

Ayres, J. M. and Clutton-Brock, T. H. (1992). River boundaries and species range size in Amazonian primates. *American Naturalist*, **140**, 531–537.

Ayres, M. P. and Scriber, J. M. (1994). Local adaptation to regional climates in *Papilio canadensis* (Lepidoptera: Papilionidae). *Ecological Monographs*, **64**, 465–482.

Bahri-Sfar, L., Lemaire, C., Hassine, O. K. B. and Bonhomme, F. (2000). Fragmentation of sea bass populations in the western and eastern Mediterranean as revealed by microsatellite polymorphism. *Proceedings of the Royal Society, London*, B **267**, 929–935.

Bahus, T. K. (1993). Clutch size variation of the Meadow Pipit *Anthus pratensis* in relation to altitude and latitude. *Fauna Norvegica Series C, Cinclus*, **16**, 37–40.

Baker, G. H. (1985). The distribution and abundance of the Portuguese millipede *Ommatoiulus moreletii* (Diplopoda: Iulidae) in Australia. *Australian Journal of Ecology*, **10**, 249–259.

Baker, M. (1995). Environmental component of latitudinal clutch-size variation in house sparrows (*Passer domesticus*). *Auk*, **112**, 249–252.

Baker, R. H. A. (1996). Developing a European pest risk mapping system. *Bulletin OEPP/EPPO Bulletin*, **26**, 485–494.

Baker, R. H. A., Cannon, R. J. C. and Walters, K. F. A. (1996). An assessment of the risks posed by selected non-indigenous pests to UK crops under climate change. *Aspects of Applied Biology*, **45**, 323–330.

Baldwin, J. D. and Dingle, H. (1986). Geographic variation in the effects of temperature in life-history traits in the large milkweed bug *Oncopeltus fasciatus*. *Oecologia*, **69**, 64–71.

Barber, P. H., Palumbi, S. R., Erdmann, M. V. and Moosa, M. K. (2000). A marine Wallace's line? *Nature*, **406**, 692–693.

Barnes, H. (1957). The northern limits of *Balanus balanoides* (L.). *Oikos*, **8**, 1–14.

Barnes, H. (1958). Regarding the southern limits of *Balanus balanoides* (L.). *Oikos*, **9**, 139–157.

Barraclough, T. G. and Vogler, A. P. (2000). Detecting the geographical pattern of speciation from species-level phylogenies. *American Naturalist*, **155**, 419–434.

Bart, J. and Klosiewski, S. P. (1989). Use of presence-absence to measure changes in avian density. *Journal of Wildlife Management*, **53**, 847–852.

Barton, N. H. and Charlesworth, B. (1984). Genetic revolutions, founder effects, and speciation. *Annual Review of Ecology and Systematics*, **15**, 133–164.

Baskauf, C. J., McCauley, D. E. and Eickmeier, W. G. (1994). Genetic analyses of a rare and a widespread species of *Echinacea* (Asteraceae). *Evolution*, **48**, 180–188.

Bateman, M. A. (1967). Adaptations to temperature in geographic races of the Queensland fruit fly, *Dacus* (*Strumeta*) *tryoni*. *Australian Journal of Zoology*, **15**, 1141–1161.

Battisti, C. and Contoli, L. (1999). Mean range size of the species, bird richness and ecogeographical factors: data from Italian peninsula and islands. *Avocetta*, **23**, 48–57.

Bazzaz, F. A. (1996). *Plants in changing environments: linking physiological, population, and community ecology.* Cambridge University Press, Cambridge.

Beard, K. H., Hengartner, N. and Skelly, D. K. (1999). Effectiveness of predicting breeding bird distibutions using probabilistic models. *Conservation Biology*, **13**, 1108–1116.

Beddington, J. R., Free, C. A. and Lawton, J. H. (1976). Concepts of stability and resilience in predator-prey models. *Journal of Animal Ecology*, **45**, 791–816.

Beerling, D. J. (1993). The impact of the temperature on the northern distribution limits of the introduced species *Fallopia japonica* and *Impatiens glandulifera* in north-west Europe. *Journal of Biogeography*, **20**, 45–53.

Beerling, D. J., Huntley, B. and Bailey, J. P. (1995). Climate and the distribution of *Fallopia japonica*: use of an introduced species to test the predictive capacity of response surfaces. *Journal of Vegetation Science*, **6**, 269–282.

Bell, C. P. (1996). The relationship between geographic variation in clutch size and migration pattern in the yellow wagtail. *Bird Study*, **43**, 333–341.

Bell, G. and Burt, A. (1991). The comparative biology of parasite species diversity: internal helminths of freshwater fish. *Journal of Animal Ecology*, **60**, 1047–1063.

Bengtsson, K. (1993). *Fumana procumbens* on Öland—population dynamics of a disjunct species at the northern limit of its range. *Journal of Ecology*, **81**, 745–758.

Bengtsson, K. (2000). Long-term demographic variation in range-margin populations of *Gypsophila fastigiata*. *Folia Geobotanica*, **35**, 143–160.

Bennett, K. D. (1997). *Evolution and ecology: the pace of life.* Cambridge University Press, Cambridge.

Bensch, S. (1999). Is the range size of migratory birds constrained by their migratory program? *Journal of Biogeography*, **26**, 1225–1235.

Bensch, S. and Hasselquist, D. (1999). Phylogeographic population structure of great reed warblers: an analysis of mtDNA control region sequences. *Biological Journal of the Linnean Society*, **66**, 171–185.

Benton, M. J. (1994). Palaeontological data and identifying mass extinctions. *Trends in Ecology and Evolution*, **9**, 181–185.

Bergeron, Y. and Brisson, J. (1990). Fire regime in red pine stands at the northern limit of the species range. *Ecology*, **71**, 1352–1364.

Bergeron, Y. and Gagnon, D. (1987). Age structure of red pine (*Pinus resinosa* Ait.) at its northern limit in Quebec. *Canadian Journal of Forestry Research*, **17**, 129–137.

Bergmann, C. (1847). Ueber die Verhältnisse der Wärmeökonomie der Thiere zu ihrer Grösse. *Gottinger Studien*, **3**, 595–708.

Bernardi, G. (2000). Barriers to gene flow in *Embiotoca jacksoni*, a marine fish lacking a pelagic larval stage. *Evolution*, **54**, 226–237.

Betancourt, J. L., Schuster, W. S., Mitton, J. B. and Anderson, R. S. (1991). Fossil and genetic history of a pinyon pine (*Pinus edulis*) isolate. *Ecology*, **72**, 1685–1697.

Bider, J. R. and Morrison, K. A. (1981). Changes in toad (*Bufo americanus*) responses to abiotic factors at the northern limit of their distribution. *American Midland Naturalist*, **106**, 293–304.

Binns, M. R. (1986). Behavioural dynamics and the negative binomial distribution. *Oikos*, **47**, 315–318.

Birch, L. C. (1953). Experimental background to the study of the distribution and abundance of insects. I. The influence of temperature, moisture and food on the innate capacity for increase of three grain beetles. *Ecology*, **34**, 698–711.

Birch, L. C., Dobzhansky, T., Elliott, P. O. and Lewontin, R. C. (1963). Relative fitness of geographic races of *Drosophila serrata*. *Evolution* **17**, 72–83.

BirdLife International (2000). *Threatened birds of the world*. Lynx Edicions and BirdLife International, Barcelona & Cambridge.

Black, R. and Prince, J. (1983). Fauna associated with the coral *Pocillopora damicornis* at the southern limit of its distribution in Western Australia. *Journal of Biogeography*, **10**, 135–152.

Blackburn, T. M. and Duncan, R. P. (2001a). Determinants of establishment success in introduced birds. *Nature*, **414**, 195–197.

Blackburn, T. M. and Duncan, R. P. (2001b). Establishment patterns of exotic birds are constrained by non-random patterns in introduction. *Journal of Biogeography*, **28**, 927–939.

Blackburn, T. M. and Gaston, K. J. (1996a). Spatial patterns in the geographic range sizes of bird species in the New World. *Philosophical Transactions of the Royal Society, London*, B **351**, 897–912.

Blackburn, T. M. and Gaston, K. J. (1996b). A sideways look at patterns in species richness, or why there are so few species outside the tropics. *Biodiversity Letters*, **3**, 44–53.

Blackburn, T. M. and Gaston, K. J. (2001). Linking patterns in macroecology. *Journal of Animal Ecology*, **70**, 338–352.

Blackburn, T. M., Gaston, K. J. and Gregory, R. D. (1997a). Abundance-range size relationships in British birds: is unexplained variation a product of life history? *Ecography*, **20**, 466–474.

Blackburn, T. M., Gaston, K. J. and Lawton, J. H. (1998a). Patterns in the geographic ranges of the world's woodpeckers. *Ibis*, **140**, 626–638.

Blackburn, T. M., Gaston, K. J., Greenwood, J. J. D. and Gregory, R. D. (1998b). The anatomy of the interspecific abundance-range size relationship for the British avifauna: II. Temporal dynamics. *Ecology Letters*, **1**, 47–55.

Blackburn, T. M., Gaston, K. J., Quinn, R. M. and Gregory, R. D. (1999). Do local abundances of British birds change with proximity to range edge? *Journal of Biogeography*, **26**, 493–505.

Blackburn, T. M., Gaston, K. J., Quinn, R. M., Arnold, H. and Gregory, R. D. (1997b). Of mice and wrens: the relation between abundance and geographic range size in British mammals and birds. *Philosophical Transactions of the Royal Society, London*, B **352**, 419–427.

Blackburn, T. M., Lawton, J. H. and Perry, J. N. (1992). A method of estimating the slope of upper bounds of plots of body size and abundance in natural animal assemblages. *Oikos*, **65**, 107–112.

Blakers, N., Davies, S. J. J. F. and Reilly, P. N. (1984). *The atlas of Australian birds*. Melbourne University Press, Carlton.

Blegvad, H. (1929). *Mortality among marine animals of the littoral region in ice waters.* Report of the Danish Biological Station to the Board of Agriculture (Ministry of Fisheries), Copenhagen, **35**, 49–62.

Blockstein, D. E. and Tordoff, H. B. (1985). A contemporary look at the extinction of the passenger pigeon. *American Birds*, **39**, 845–851.

Blows, M. W. (1993). The genetics of central and marginal populations of *Drosophila serrata*. II. Hybrid breakdown in fitness components as a correlated response to selection for desiccation resistance. *Evolution*, **47**, 1271–1285.

Blows, M. W. and Hoffmann, A. A. (1993). The genetics of central and marginal populations of *Drosophila serrata*. I. Genetic variation for stress resistance and species borders. *Evolution*, **47**, 1255–1270.

Bock, C. E. (1984). Geographical correlates of abundance vs. rarity in some Noth American winter landbirds. *The Auk*, **101**, 266–273.

Bock, C. E. (1987). Distribution-abundance relationships of some Arizona landbirds: a matter of scale? *Ecology*, **68**, 124–129.

Bock, C. E., Bock, J. H. and Lepthien, L. W. (1977). Abundance patterns of some bird species wintering on the Great Plains of the U. S. A. *Journal of Biogeography*, **4**, 101–110.

Bock, C. E. and Ricklefs, R. E. (1983). Range size and local abundance of some North American songbirds: a positive correlation. *American Naturalist*, **122**, 295–299.

Boecken, B. and Shachak, M. (1998). The dynamics of abundance and incidence of annual plant species richness during colonization in a desert. *Ecography*, **21**, 63–73.

Böhning-Gaese, K., González-Guzmán, L. I. and Brown, J. H. (1998). Constraints on dispersal and the evolution of the avifauna of the Northern Hemisphere. *Evolutionary Ecology*, **12**, 767–783.

Böhning-Gaese, K., Halbe, B., Lemoine, N. and Oberrath, R. (2000). Factors influencing the clutch size, number of broods and annual fecundity of North American and European land birds. *Evolutionary Ecology Research*, **2**, 823–839.

Böhning-Gaese, K. and Oberrath, R. (1999). Phylogenetic effects on morphological, life-history, behavioural and ecological traits of birds. *Evolutionary Ecology Research*, **1**, 347–364.

Bonabeau, E., Dagorn, L. and Fréon, P. (1999). Scaling in animal group-size distributions. *Proceedings of the National Academy of Science of the USA*, **96**, 4472–4477.

Bonan, G. B. (1999). Frost followed the plow: impacts of deforestation on the climate of the United States. *Ecological Applications*, **9**, 1305–1315.

Bonan, G. B. and Sirois, L. (1992). Air temperature, tree growth, and the northern and southern range limits to *Picea mariana*. *Journal of Vegetation Science*, **3**, 495–506.

Boone, R. B. and Krohn, W. B. (2000). Relationship between avian range limits and plant transition zones in Maine. *Journal of Biogeography*, **27**, 471–482.

Booth, T. H., Nix, H. A., Hutchinson, M. F. and Jovanovic, T. (1988). Niche analysis and tree species introduction. *Forest Ecology and Management*, **23**, 47–59.

Borcard, D., Legendre, P. and Drapeau, P. (1992). Partialling out the spatial component of ecological variation. *Ecology*, **73**, 1045–1055.

Bost, C. A. and Jouventin, P. (1991). The breeding performance of the Gentoo Penguin *Pygoscelis papua* at the northern edge of its range. *Ibis*, **133**, 14–25.

Boswell, M. T. and Patil, G. P. (1970). Chance mechanisms generating the negative binomial distribution. In *Random counts in scientific work*, Vol 1 (ed. G. P. Patil), pp. 3–22. Pennsylvania State University Press, University Park.

Bowers, M. A. and Brown, J. H. (1982). Body size and coexistence in desert rodents: chance or community structure? *Ecology*, **63**, 391–400.

Boyce, M. S. (1979). Seasonality and patterns of natural selection for life histories. *American Naturalist*, **114**, 569–583.

Boyko, H. (1947). On the role of plants as quantitative climate indicators and the geo-ecological law of distribution. *Journal of Ecology*, **35**, 138–157.

Bradford, M. J., Taylor, G. C. and Allan, J. A. (1997). Empirical review of coho salmon smolt abundance and the prediction of smolt production at the regional level. *Transactions of the American Fisheries Society*, **126**, 49–64.

Bradshaw, W. E., Fujiyama, S. and Holzapfel, C. M. (2000). Adaptation to the thermal climate of North America by the pitcher-plant mosquito, *Wyeomyia smithii*. *Ecology*, **81**, 1262–1272.

Bradshaw, W. E. and Holzapfel, C. M. (2001). Genetic shift in photoperiodic response correlated with global warming. *Proceedings of the National Academy of Sciences of the USA*, **98**, 14509–14511.

Braithwaite, L. W. (1983). Studies of the arboreal marsupial fauna of eucalypt forests being harvested for woodpulp at Eden, New South Wales. I. The species and distribution of animals. *Australian Wildlife Research*, **10**, 219–229.

Braithwaite, L. W., Binns, D. L. and Nowlan, R. D. (1988). The distribution of arboreal marsupials in relation to eucalypt forest types in the Eden (NSW) Woodchip Concession Area. *Australian Wildlife Research*, **15**, 363–373.

Braithwaite, L. W., Dudzinski, M. L. and Turner, J. (1983). Studies of the arboreal marsupial fauna of eucalypt forests being harvested for woodpulp at Eden, New South Wales. II. Relationship between the fauna density, richness and diversity and measured variables of habitat. *Australian Wildlife Research*, **10**, 231–247.

Braithwaite, L. W., Turner, J. and Kelly, J. (1984). Studies of the arboreal marsupial fauna of eucalypt forests being harvested for woodpulp at Eden, New South Wales. III. Relationships between faunal densities, eucalypt occurrence and foliage nutrients and soil parent materials. *Australian Wildlife Research*, **11**, 41–48.

Branning, D. W. (eds) (1993). *Atlas of breeding birds in Pennsylvania*. University of Pittsburgh Press, Pittsburgh.

Brattstrom, B. H. (1968). Thermal acclimation in anuran amphibians as a function of latitude and altitude. *Comparative Biochemistry and Physiology*, **24**, 93–111.

Brereton, R., Bennett, S. and Mansergh, I. (1995). Enhanced greenhouse climate change and its potential effect on selected fauna of south-eastern Australia: a trend analysis. *Biological Conservation*, **72**, 339–354.

Brett, J. R. (1956). Some principles in the thermal requirements of fishes. *Quarterly Review of Biology*, **31**, 75–87.

Brewer, A. M. and Gaston, K. J. (2002). The geographical range structure of the holly leaf-miner. I. Population density. *Journal of Animal Ecology*, **71**, 99–111.

Brewer, A. M. and Gaston, K. J. (in press). The geographical range structure of the holly leaf-miner. II. Demographic rates. *Journal of Animal Ecology*.

Brewer, R., McPeek, G. A. and Adams, R. J. (1991). *The atlas of breeding birds of Michigan*. Michigan State University Press, East Lansing.

Brönmark, C. and Edenhamn, P. (1994). Does the presence of fish affect the distribution of tree frogs (*Hyla arborea*)? *Conservation Biology*, **8**, 841–845.

Brooks, D. R. and McLennan, D. A. (1991). *Phylogeny, ecology, and behavior: a research program in comparative biology*. University of Chicago Press.

Brooks, D. R. and McLennan, D. A. (1993). Comparative study of adaptive radiations with an example using parasitic flatworms (Platyhelminthes: Cercomeria). *American Naturalist*, **142**, 755–778.

Brooks, T., Balmford, A., Burgess, N., Fjeldså, J., Hansen, L. A., Moore, J., Rahbek, C. and Williams, P. (2001). Toward a blueprint for conservation in Africa. *BioScience*, **51**, 613–624.

Brown, A. H. D. and Briggs, J. D. (1991). Sampling strategies for genetic variation in *ex situ* collections of endangered plant species. In *Genetics and conservation of rare plants* (eds D. A. Falk and K. E. Holsinger), pp. 99–119. Oxford University Press, New York.

Brown, B. E. and Suharson, O. (1990). Damage and recovery of coral reefs affect by El Niño related seawater warming in the Thousand Islands, Indonesia. *Coral Reefs*, **8**, 163–170.

Brown, J. H. (1981). Two decades of homage to Santa Rosalia: toward a general theory of diversity. *American Zoologist*, **21**, 877–888.

Brown, J. H. (1984). On the relationship between abundance and distribution of species. *American Naturalist*, **124**, 255–279.

Brown, J. H. (1987). Variation in desert rodent guilds: patterns, processes, and scales. In *Organization of communities: past and present* (eds J. H. R. Gee and P. S. Giller), pp. 185–203. Blackwell Scientific, Oxford.

Brown, J. H. (1995). *Macroecology*. University of Chicago Press, Chicago.

Brown, J. H. and Kodric-Brown, A. (1977). Turnover rates in insular biogeography: effect of immigration on extinction. *Ecology*, **58**, 445–449.

Brown, J. H. and Lomolino, M. V. (1998). *Biogeography*, 2nd edn. Sinauer Associates, Sunderland, Massachusetts.

Brown, J. H. and Maurer, B. A. (1987). Evolution of species assemblages: effects of energetic constraints and species dynamics on the diversification of the American avifauna. *American Naturalist*, **130**, 1–17.

Brown, J. H. and Maurer, B. A. (1989). Macroecology: the division of food and space among species on continents. *Science*, **243**, 1145–1150.

Brown, J. H., Mehlman, D. W. and Stevens, G. C. (1995). Spatial variation in abundance. *Ecology*, **76**, 2028–2043.

Brown, J. H. and Nicoletto, P. F. (1991). Spatial scaling of species composition: body masses of North American land mammals. *American Naturalist*, **138**, 1478–1512.

Brown, J. H., Stevens, G. C. and Kaufman, D. M. (1996). The geographic range: size, shape, boundaries and internal structure. *Annual Review of Ecology and Systematics*, **27**, 597–623.

Bruelheide, H. and Scheidel, U. (1999). Slug herbivory as a limiting factor for the geographic range of *Arnica montana*. *Journal of Ecology*, **87**, 839–848.

Brussard, P. F. (1984). Geographic patterns and environmental gradients: the central-marginal model in *Drosophila* revisited. *Annual Review of Ecology and Systematics*, **15**, 25–64.

Bryant, S. R., Thomas, C. D. and Bale, J. S. (1997). Nettle-feeding nymphalid butterflies: temperature, development and distribution. *Ecological Entomology*, **22**, 390–398.

Bucher, E. H. (1992). The causes of extinction of the passenger pigeon. In *Current ornithology* (ed. D. M. Power), pp. 1–36. Plenum Press, New York.

Budd, A. F. and Coates, A. G. (1992). Nonprogressive evolution in a clade of Cretaceous Montastraea-like corals. *Paleobiology*, **18**, 425–446.

Bullock, J. M., Edwards, R. J., Carey, P. D. and Rose, R. J. (2000). Geographical separation of two *Ulex* species at three spatial scales: does competition limit species' ranges? *Ecography*, **23**, 257–271.

Burban, C., Petit, R. J., Carcreff, E. and Jactel, H. (1999). Rangewide variation of the maritime pine bast scale *Matsucoccus feytaudi* Duc. (Homoptera: Matsucoccidae) in relation to the genetic structure of its host. *Molecular Ecology*, **8**, 1593–1602.

Burgman, M. A. (1989). The habitat volumes of scarce and ubiquitous plants: a test of the model of environmental control. *American Naturalist*, **133**, 228–239.

Burgman, M. A. and Lindenmayer, D. B. (1998). *Conservation biology for the Australian environment*. Surrey Beatty, Chipping Norton.

Burke, M. J., Gusta, L. V., Quamme, H. A., Weiser, C. J. and Li, P. H. (1976). Freezing and injury in plants. *Annual Review of Plant Physiology*, **27**, 507–528.

Burton, J. F. (1995). *Birds and climate change*. London, Christopher Helm.

Burton, J. F. (2001). The response of European insects to climate change. *British Wildlife*, **12**, 188–198.

Bush, G. L. (1975). Modes of animal speciation. *Annual Review of Ecology and Systematics*, **6**, 334–364.

Bush, M. B. (1994). Amazonian speciation: a necessarily complex model. *Journal of Biogeography*, **21**, 5–17.

Bustamante, J. (1997). Predictive models for lesser kestrel *Falco naumanni* distribution, abundance and extinction in southern Spain. *Biological Conservation*, **80**, 153–160.

Buzas, M. A. and Culver, S. J. (1991). Species diversity and dispersal of benthic foraminifera. *BioScience*, **41**, 483–489.

Buzas, M. A. and Culver, S. J. (1999). Understanding regional species diversity through the log series distribution of occurrences. *Diversity and Distributions*, **8**, 187–195.

Buzas, M. A., Koch, C. F., Culver, S. J. and Sohl, N. F. (1982). On the distribution of species occurrence. *Paleobiology*, **8**, 143–150.

Cade, T. J. and Woods, C. P. (1997). Changes in distribution and abundance of the loggerhead shrike. *Conservation Biology*, **11**, 21–31.

Cadman, M. D., Eagles, P. F. J. and Helleiner, F. M. (1987). *The atlas of breeding birds of Ontario*. Federation of Ontario Naturalists, Don Mills, Ontario.

Cambefort, Y. (1994). Body size, abundance, and geographical distribution of Afrotropical dung beetles (Coleoptera: Scarabaeidae). *Acta Oecologica*, **15**, 165–179.

Cambridge, M. L., Breeman, A. M. and van den Hoek, C. (1990). Temperature limits at the distribution boundaries of four tropical to temperate species of *Cladophora* (Cladophorales: Chlorophyta) in the North Atlantic Ocean. *Aquatic Botany*, **38**, 135–151.

Cameron, R. A. D. (1998). Dilemmas of rarity: biogeographical insights and conservation priorities for land Mollusca. *Journal of Conchology Special Publication*, **2**, 51–60.

Cammell, M. E., Tatchell, G. M. and Woiwod, I. P. (1989). Spatial pattern of abundance of the black bean aphid, *Aphis fabae*, in Britain. *Journal of Applied Ecology*, **26**, 463–472.

Campbell, B. and Lack, E. (eds) (1985). *A dictionary of birds*. Poyser, Calton.

Cannon, R. J. C. (1998). The implications of predicted climate change for insect pests in the UK, with emphasis on non-indigenous species. *Global Change Biology*, **4**, 785–796.

Cantor, L. F. and Whitham, T. G. (1989). Importance of belowground herbivory: pocket gophers may limit aspen to rock outcrop refugia. *Ecology*, **70**, 962–970.

Capparella, A. P. (1991). Neotropical avian diversity and riverine barriers. *Acta XX Congressus Internationalis Ornithologici*, 307–316.

Carbon Dioxide Information Analysis Centre (2000). Current greenhouse gas concentrations. http://cdiac.esd.orul.gov/pns/current_ghg.html

Carey, P. D. (1996). DISPERSE: A cellular automaton for predicting the distribution of species in a changed climate. *Global Ecology and Biogeography Letters*, **5**, 217–226.

Carey, P. D., Watkinson, A. R. and Gerard, F. F. O. (1995). The determinants of the distribution and abundance of the winter annual grass *Vulpia ciliata* ssp. *ambigua*. *Journal of Ecology*, **83**, 177–187.

Carlton, J. T., Geller, J. B., Reaka-Kudla, M. L. and Norse, E. A. (1999). Historical extinctions in the sea. *Annual Review of Ecology and Systematics*, **30**, 515–538.

Carne, P. B. (1965). Distribution of the eucalypyt-defoliating sawfly *Perga affinis affinis* (Hymenoptera). *Australian Journal of Zoology*, **13**, 593–612.

Carrascal, L. M., Bautista, L. M. and Lázaro, E. (1993). Geographical variation in the density of the white stork *Ciconia ciconia* in Spain: influence of habitat structure and climate. *Biological Conservation*, **65**, 83–87.

Carroll, S. S. and Pearson, D. L. (2000). Detecting and modelling spatial and temporal dependence in conservation biology. *Conservation Biology*, **14**, 1893–1897.

Carson, H. L. (1959). Genetic conditions which promote or retard the formation of species. *Cold Spring Harbor Symposium in Quantitative Biology*, **24**, 87–105.

Carter, R. N. and Prince, S. D. (1981). Epidemic models used to explain biogeographical distribution limits. *Nature*, **293**, 644–645.

Carter, R. N. and Prince, S. D. (1985a). The geographical distribution of prickly lettuce (*Lactuca serriola*). I. A general survey of its habitats and performance in Britain. *Journal of Ecology*, **73**, 27–38.

Carter, R. N. and Prince, S. D. (1985b). The effect of climate on plant distributions. In *The climatic scene* (eds M. J. Tooley and J. Sheail), pp. 235–254. Allen and Unwin, London.

Carter, R. N. and Prince, S. D. (1988). Distribution limits from a demographic viewpoint. In *Plant population ecology* (eds A. J. Davy, M. J. Hutchings and A. R. Watkinson), pp. 165–184. Blackwell Scientific, Oxford.

Carter, T. R., Parry, M. L. and Porter, J. H. (1991a). Climatic change and future agroclimatic potential in Europe. *International Journal of Climatology*, **11**, 251–269.

Carter, T. R., Porter, J. H. and Parry, M. L. (1991b). Climatic warming and crop potential in Europe. *Global Environmental Change*, **1**, 291–312.

Carthew, S. M., Goldingay, R. L. and Funnell, D. L. (1999). Feeding behaviour of the yellow-bellied glider (*Petaurus australis*) at the western edge of its range. *Wildlife Research*, **26**, 199–208.

Case, T. J. and Taper, M. L. (2000). Interspecific competition, environmental gradients, gene flow, and the coevolution of species' borders. *American Naturalist*, **155**, 583–605.

Casey, J. M. and Myers, R. A. (1998). Near extinction of a large, widely distributed fish. *Science*, **281**, 690–692.

Catterall, C. P., Kingston, M. B., Park, K. and Sewell, S. (1998). Deforestation, urbanisation and seasonality: interacting effects on a regional bird assemblage. *Biological Conservation*, **84**, 65–81.

Caughley, G. (1970). Eruption of ungulate populations, with emphasis on Himalayan thar in New Zealand. *Ecology*, **51**, 53–72.

Caughley, G., Grice, D., Braker, R. and Brown, B. (1988). The edge of the range. *Journal of Animal Ecology*, **57**, 771–785.

Caughley, G., Short, J., Grigg, G. C. and Nix, H. (1987). Kangaroos and climate: an analysis of distribution. *Journal of Animal Ecology*, **56**, 751–761.

Cawthorne, R. A. and Marchant, J. H. (1980). The effects of the 1978/79 winter on British bird populations. *Bird Study*, **27**, 163–172.

Ceballos, G., Rodríguez, P. and Medellín, R. A. (1998). Assessing conservation priorities in megadiverse Mexico: mammalian diversity, endemicity, and endangerment. *Ecological Applications*, **8**, 8–17.

Cerrano, C., Bavestrello, G., Bianchi, C. N., Cattaneo-vietti, R., Bava, S., Morganti, C., Morri, C., Picco, P., Sara, G., Schiaparelli, S., Siccardi, A. and Sponga, F. (2000). A catastrophic mass-mortality episode of gorgonians and other organisms in the Ligurian Sea (North-western Mediterranean), summer 1999. *Ecology Letters*, **3**, 284–293.

Channell, R. and Lomolino, M. V. (2000a). Dynamic biogeography and conservation of endangered species. *Nature*, **403**, 84–86.

Channell, R. and Lomolino, M. V. (2000b). Trajectories to extinction: spatial dynamics of the contraction of geographical ranges. *Journal of Biogeography*, **27**, 169–179.

Chapin, F. S. III and Chapin, M. C. (1981). Ecotypic differentiation of growth processes in *Carex aquatilis* along latitudinal and local gradients. *Ecology*, **62**, 1000–1009.

Chesser, R. T. and Zink, R. M. (1994). Modes of speciation in birds: a test of Lynch's method. *Evolution*, **48**, 490–497.

Chiang, H. C. (1961). Fringe populations of the European corn borer *Pyrausta nubilalis*: their characteristics and problems. *Annals of the Entomological Society of America*, **54**, 378–387.

Chown, S. L. (1997). Speciation and rarity: separating cause from consequence. In *The biology of rarity: causes and consequences of rare-common differences* (eds W. E. Kunin and K. J. Gaston), pp. 91–109. Chapman and Hall, London.

Chown, S. L. and Gaston, K. J. (2000). Areas, cradles, and museums: the latitudinal gradient in species richness. *Trends in Ecology and Evolution*, **15**, 311–315.

Chown, S. L., Gaston, K. J. and Williams, P. H. (1998). Global patterns in species richness of pelagic seabirds: the Procellariiformes. *Ecography*, **21**, 342–350.

Chown, S. L., Rodrigues, A. S. L., Gremmen, N. J. M. and Gaston, K. J. (2001). World Heritage status and conservation of southern ocean islands. *Conservation Biology*, **15**, 550–557.

Christiansen, M. B. and Pitter, E. (1997). Species loss in a forest bird community near Logoa Santa in southeastern Brazil. *Biological Conservation*, **80**, 23–32.

Chung, M. G. and Chung, M. Y. (2000). Levels and partitioning of genetic diversity of *Camellia japonica* (Theaceae) in Korea and Japan. *Silvae Genetica*, **49**, 119–124.

Claridge, M. F. and Wilson, M. R. (1982). Insect herbivore guilds and species area relationships: leafminers on British trees. *Ecological Entomology*, **7**, 19–30.

Clarke, A. and Lidgard, S. (2000). Spatial patterns of diversity in the sea: bryozoan species richness in the North Atlantic. *Journal of Animal Ecology*, **69**, 799–814.

Clay, K., Dement, D. and Rejmanek, M. (1985). Experimental evidence for host races in mistletoe (*Phoradendron tomentosum*). *American Journal of Botany*, **72**, 1225–1231.

Clutton-Brock, T. H. and Harvey, P. H. (1977). Primate ecology and social organisation. *Journal of Zoology, London*, **183**, 1–39.

Cody, M. L. (1966). A general theory of clutch size. *Evolution*, **20**, 174–184.

Cody, M. L. and Overton, J. M. (1996). Short-term evolution of reduced dispersal in island plant populations. *Journal of Ecology*, **84**, 53–61.

Collar, N. J., Crosby, M. J. and Stattersfield, A. J. (1994). *Birds to watch 2. The world list of threatened birds*. BirdLife International, Cambridge.

Collins, S. L. and Glenn, S. M. (1990). A hierarchical analysis of species' abundance patterns in grassland vegetation. *American Naturalist*, **135**, 633–648.

Collins, S. L. and Glenn, S. M. (1997). Effects of organismal and distance scaling on analysis of species distribution and abundance. *Ecological Applications*, **7**, 543–551.

Colwell, R. K. and Hurtt, G. C. (1994). Nonbiological gradients in species richness and a spurious Rapoport effect. *American Naturalist*, **144**, 570–595.

Condit, R., Ashton, P. S., Baker, P., Bunyavejchewin, S., Gunatilleke, S., Gunatilleke, N., Hubbell, S. P., Foster, R. B., Itoh, A., LaFrankie, J. V., Seng Lee, H., Losos, E., Manokaran, N., Sukumar, R. and Yamakura, T. (2000). Spatial patterns in the distribution of tropical tree species. *Science*, **288**, 1414–1418.

Conkey, L. E., Keifer, M. and Lloyd, A. H. (1995). Disjunct jack pine (*Pinus banksiana* Lamb) structure and dynamics, Acadia National Park, Maine. *Écoscience*, **2**, 168–176.

Connor, E. F. and Bowers, M. A. (1987). The spatial consequences of interspecific competition. *Annales Zoologici Fennici*, **24**, 213–226.

Conquillat, M. (1951). Sur les plantes les plus communes a la surface du globe. *Bulletin Mensuel Société Botanique de Lyon*, **20**, 165–170.

Conrad, K. F., Perry, J. N. and Woiwod, I. P. (2001). An abundance-occupancy time-lag during the decline of an arctiid tiger moth. *Ecology Letters*, **4**, 300–303.

Cook, W. C. (1924). The distribution of the pale western cutworm, *Porosagrotis orthogonia* Morr.: a study in physical ecology. *Ecology*, **5**, 60–69.

Coope, G. R. (1978). Constancy of insect species versus inconstancy of Quaternary environments. In *Diversity of insect faunas* (eds L. A. Mound and N. Waloff), pp. 176–187. Blackwell Scientific, Oxford.

Coope, G. R. and Lemdahl, G. (1996). Validations for the use of beetle remains as reliable indicators of Quaternary climates: a reply to the criticisms by Johan Andersen. *Journal of Biogeography*, **23**, 115–121.

Coope, G. R., Lemdahl, G., Lowe, J. J. and Walkling, A. (1998). Temperature gradients in northern Europe during the last glacial-Holocene transition (14–9 C-14 kyr BP). *Journal of Quaternary Science*, **13**, 419–433.

Corbacho, C., Sánchez, J. M. and Sánchez, A. (1997). Breeding biology of Montagu's Harrier *Circus pygargus* L. in agricultural environments of southwest Spain; comparison with other populations in the western Palearctic. *Bird Study*, **44**, 166–175.

Corbet, G. B. and Harris, S. (eds) (1991). *The handbook of British mammals*. Blackwell Scientific, Oxford.

Cornelius, C., Cofré, H. and Marquet, P. A. (2000). Effects of habitat fragmentation on bird species in a relict temperate forest in semiarid Chile. *Conservation Biology*, **14**, 534–543.

Cornell, H. V. and Washburn, J. O. (1979). Evolution of the richness area correlation for cynipid gall wasps on oak trees: a comparison of two geographic areas. *Evolution*, **33**, 257–274.

Coulson, J. C. and Whittaker, J. B. (1978). Ecology of moorland animals. In *Ecological studies 27. Production ecology of British moors and montane grasslands* (eds Heal, O. W. and Perkins, D. F.), pp. 52–93. Springer-Verlag, Berlin.

Cousens, R. and Mortimer, M. (1995). *Dynamics of weed populations*. Cambridge University Press, Cambridge.

Cowen, R. K., Lwiza, K. M. M., Sponaugle, S., Paris, C. B. and Olson, D. B. (2000). Connectivity of marine populations: open or closed? *Science*, **287**, 857–859.

Cowley, M. J. R., Thomas, C. D., Thomas, J. A. and Warren, M. S. (1999). Flight areas of British butterflies: assessing species status and decline. *Proceedings of the Royal Society, London*, B **266**, 1587–1592.

Cowley, M. J. R., Thomas, C. D., Roy, D. B., Wilson, R. J., León-Cortés, J. L., Gutiérrez, D., Bulman, C. R., Quinn, R. M., Moss, D. and Gaston, K. J. (2001). Density-distribution relationships in British butterflies I. The effect of mobility and spatial scale. *Journal of Animal Ecology*, **70**, 410–425.

Cowlishaw, G. and Hacker, J. E. (1997). Distribution, diversity, and latitude in African primates. *American Naturalist*, **150**, 505–512.

Coyle, B. F., Sharik, T. L. and Feret, P. P. (1982). Variation in leaf morphology and disjunct and continuous populations of river birch (*Betula nigra* L.). *Silvae Genetica*, **31**, 122–125.

Cracraft, J. (1982). Geographic differentiation, cladistics, and vicariance biogeography: reconstructing the tempo and mode of evolution. *American Zoologist*, **22**, 411–424.

Cracraft, J. (1986). Origin and evolution of continental biotas: speciation and historical congruence within the Australian avifauna. *Evolution*, **40**, 977–996.

Cracraft, J. and Prum, R. O. (1988). Patterns and processes of diversification: speciation and historical congruence in some neotropical birds. *Evolution*, **42**, 603–620.

Crame, J. A. (1993). Bipolar molluscs and their evolutionary implications. *Journal of Biogeography*, **20**, 145–161.

Cramp, S. (ed.) (1977). *Handbook of the birds of Europe, the Middle East and North Africa. The Birds of the western Palearctic. Vol. 1. Ostrich to Ducks*. Oxford University Press, Oxford.

Crawley, M. J. (1987). What makes a community invasible? In *Colonisation, succession and stability* (eds A. J. Gray, M. J. Crawley and P. J. Edwards), pp. 429–453. Blackwell Scientific, Oxford.

Crecco, V. and Overholtz, W. J. (1990). Causes of density-dependent catchability for Georges Bank haddock *Melanogrammus aeglefinus*. *Canadian Journal of Fisheries and Aquatic Sciences*, **47**, 385–394.

Crisp, D. J. (1965). Observations on the effects of climate and weather on marine communities. In *The biological significance of climatic changes in Britain* (eds C. G. Johnson and L. P. Smith), pp. 63–77. Academic Press, London.

Crowson, R. A. (1981). *The biology of the Coleoptera*. Academic Press, London.

Cumming, G. S. (2000). Using between-model comparisons to fine-tune linear models of species ranges. *Journal of Biogeography*, **27**, 441–455.

Curio, E. (1989). Some aspects of avian mortality patterns. *Mitteilungen aus dem Zoologischen Museum in Berlin*, **65** (*Supplementband Annalen Fuer Ornithologie* 13), 47–70.

Curnutt, J. L., Pimm, S. L. and Maurer, B. A. (1996). Population variability of sparrows in space and time. *Oikos*, **76**, 131–144.

Curran, P. J., Foody, G. M. and van Gardingen, P. R. (1997). Scaling-up. In *Scaling-up: from cell to landscape* (eds. P. R., van Gardingen, G. M. Foody and P. J. Curran), pp. 1–5. Cambridge University Press, Cambridge.

Cwynar, L. C. and MacDonald, G. M. (1987). Geographical variation of lodgepole pine in relation to population history. *American Naturalist*, **129**, 463–469.

Cyrus, D. and Robson, N. (1980). *Bird atlas of Natal*. University of Natal Press, Scottsville.

Dahl, E. (1951). On the relation between Summer temperature and the distribution of alpine vascular plants in the lowlands of Fennoscandia. *Oikos*, **3**, 22–52.

Dahl, E. (1992). Relations between macro-meteorological factors and the distribution of vascular plants in northern Europe. *Universitetet i Trondheim Vitenskapsmuseet Rapport Botanisk Serie*, 1992–1, 31–59.

Damuth, J. (1987). Interspecific allometry of population density in mammals and other animals: the independence of body mass and population energy use. *Biological Journal of the Linnean Society*, **31**, 193–246.

Dandova, R., Weidinger, K. and Zavadil, V. (1998). Morphometric variation, sexual size dimorphism and character scaling in a marginal population of Montandon's newt *Triturus montandoni* from the Czech Republic. *Italian Journal of Zoology*, **65**, 399–405.

Daniels, R. J. R., Hegde, M., Joshi, N. V. and Gadgil, M. (1991). Assigning conservation value: a case study from India. *Conservation Biology*, **5**, 464–475.

Darwin, C. (1856–1858) [publ. 1975]. *Charles Darwin's natural selection: being the second part of his big species book written from 1856–1858* (ed. R. C. Stauffer). Cambridge University Press, Cambridge.

Darwin, C. (1859). *On the origin of species by means of natural selection, or the preservation of favoured races in the struggle for life.* John Murray, London.

Darwin, C. (1888). *The formation of vegetable mould through the action of worms.* John Murray, London.

Daubenmire, R. (1985). The western limits of the range of the American bison. *Ecology*, **66**, 622–624.

Daugherty, C. H., Gibbs, G. W. and Hitchmough, R. A. (1993). Mega-island or micro-continent? New Zealand and its fauna. *Trends in Ecology and Evolution*, **8**, 437–442.

Davies, K. I. and Margules, C. R. (1998). Effects of habitat fragmentation on carabid beetles: experimental evidence. *Journal of Animal Ecology*, **67**, 460–471.

Davis, A. J., Jenkinson, L. S., Lawton, J. H., Shorrocks, B. and Wood, S. (1998a). Making mistakes when predicting shifts in species range in response to global warming. *Nature*, **391**, 783–786.

Davis, A. J., Lawton, J. H., Shorrocks, B. and Jenkinson, L. S. (1998b). Individualistic species responses invalidate simple physiological models of community dynamics under global environmental change. *Journal of Animal Ecology*, **67**, 600–612.

Davison, A. W. (1970). The ecology of *Hordeum murinum* L. I. Analysis of the distribution in Britain. *Journal of Ecology*, **58**, 453–466.

Davison, A. W. (1977). The ecology of *Hordeum murinum* L. III. Some effects of adverse climate. *Journal of Ecology*, **65**, 523–530.

De'Ath, G. (1999). Principal curves: a new technique for indirect and direct gradient analysis. *Ecology*, **80**, 2237–2253.

Dekker, R. and Beukema, J. J. (1993). Dynamics and growth of a bivalve, *Abra tenuis*, at the northern edge of its distribution. *Journal of the Marine Biological Association*, **73**, 497–511.

Dekker, R. W. R. J. (1989). Predation and the western limits of megapode distribution (Megapodiidae; Aves). *Journal of Biogeography*, **16**, 317–321.

de la Ville, N., Cousins, S. and Bird, C. (1998). Habitat suitability analysis using logistic regression and GIS to outline potential areas for conservation of the grey wolf (*Canis lupus*). In *Innovations in GIS 5* (ed. S. Carver), pp. 187–197. Taylor and Francis Group, London.

del Hoyo, J., Elliott, A. and Sargatal, J. (eds) (1992). *Handbook of the birds of the world.* Vol. 1. Lynx Edicions, Barcelona.

del Hoyo, J., Elliott, A. and Sargatal, J. (eds) (1994). *Handbook of the birds of the world.* Vol. 2. Lynx Edicions, Barcelona.

Dennis, M. K. (1996). *Tetrad atlas of the breeding birds of Essex.* The Essex Birdwatching Society, Colchester.

Dennis, R. L. H. (1993). *Butterflies and climate change.* Manchester University Press, Manchester.

Dennis, R. L. H., Donato, B., Sparks, T. H. and Pollard, E. (2000). Ecological correlates of island incidence and geographical range among British butterflies. *Biodiversity and Conservation*, **9**, 343–359.

Denys, C. and Schmidt, H. (1998). Insect communities on experimental mugwort (*Artemisia vulgaris* L.) plots along an urban gradient. *Oecologia*, **113**, 269–277.

DeSante, D. F., Burton, K. M., Saracco, J. F. and Walker, B. L. (1995). Productivity indices and survival rate estimates from MAPS, a continent-wide programme of constant effort mist-netting in North America. *Journal of Applied Statistics*, **22**, 935–947.

Descimon, H. and Napolitano, M. (1993). Enzyme polymorphism, wing pattern variability, and gepographical isolation in an endangered butterfly species. *Biological Conservation*, **66**, 117–123.

Desmarchelier, J. M. (1988). The relationship between wet-bulb temperature and the intrinsic rate of increase of eight species of stored-product Coleoptera. *Journal of Stored Product Research*, **24**, 107–113.

Despland, E. and Houle, G. (1997). Climate influences on growth and reproduction of *Pinus banksiana* (Pinaceae) at the limit of the species distribution in eastern North America. *American Journal of Botany*, **84**, 928–937.

Desponts, M. and Payette, S. (1992). Recent dynamics of jack pine at its northern distribution limit in northern Quebec. *Canadian Journal of Botany*, **70**, 1157–1167.

Diamond, J. M. (1984). 'Normal' extinctions of isolated populations. In *Extinctions* (eds M. H. Nitecki), pp. 191–246. University of Chicago Press, Chicago.

Dias, P. C. (1996). Sources and sinks in population biology. *Trends in Ecology and Evolution*, **11**, 326–330.

Díaz, M., Carbonell, R., Santos, T. and Tellería, J. L. (1998). Breeding bird communities in pine plantations of the Spanish plateaux: biogeography, landscape and vegetation effects. *Journal of Applied Ecology*, **35**, 562–574.

Diekmann, M. and Lawesson, J. E. (1999). Shifts in ecological behaviour of herbaceous forest species along a transect from northern Central to North Europe. *Folia Geobotanica*, **34**, 127–141.

Dillon, L. S. (1966).The life cycle of the species: an extension of current concepts. *Systematic Zoology*, **15**, 112–126.

Dobzhansky, T. (1950). Evolution in the tropics. *American Scientist*, **38**, 209–221.

Donald, P. F. and Fuller, R. J. (1998). Ornithological atlas data: a review of uses and limitations. *Bird Study*, **45**, 129–145.

Dow, D. D. (1969). Home range and habitat of the cardinal in peripheral and central populations. *Canadian Journal of Zoology*, **47**, 103–114.

Duncan, R. P., Bomford, M., Forsyth, D. M. and Conibear, L. (2001). High predictability in introduction outcomes and the geographical range size of introduced Australian birds: a role for climate. *Journal of Animal Ecology*, **70**, 621–632.

Dunn, P. O., Thusius, K. J., Kimber, K. and Winkler, D. W. (2000). Geographic and ecological variation in clutch size of tree swallows. *Auk*, **117**, 215–221.

Dunning, J. B. Jr (1984). Body weights of 686 species of North American birds. *Western Bird Banding Association Monograph* 1.

Dunning, J. B. (1992). *CRC handbook of avian body masses.* CRC Press, Boca Raton, Florida.

Durka, W. (1999). Genetic diversity in peripheral and subcentral populations of *Corrigiola litoralis* L. (Illecebraceae). *Heredity*, **83**, 476–484.

Durrer, S. and Schmid-Hempel, P. (1995). Parasites and the regional distribution of bumblebee species. *Ecography*, **18**, 114–122.

Ebenhard, T. (1988). Introduced birds and mammals and their ecological effects. *Swedish Wildlife Research*, **13**, 1–107.

Eckert, C. G. and Barrett, S. C. H. (1993). Clonal reproduction and patterns of genotypic diversity in *Decodon verticillatus* (Lythraceae). *American Journal of Botany*, **80**, 1175–1182.

Edenhamn, P., Hoggren, M. and Carlson, A. (2000). Genetic diversity and fitness in peripheral and central populations of the European tree frog *Hyla arborea*. *Heriditas*, **133**, 115–122.

Edwards, W. and Westoby, M. (1996). Reserve mass and dispersal investment in relation to geographic range of plant species: phylogenetically independent contrasts. *Journal of Biogeography*, **23**, 329–338.

Eeley, H. A. C. (1994). A digital method for the analysis of primate range boundaries. In *Current primatology*, Vol. 1 (eds B. Thierry, J. R. Anderson, J. J. Roeder and N. Herrenschmidt), pp. 123–132. Selected Proceedings of the XIVth Congress of the International Primatological Society, Strasbourg, France.

Eeley, H. A. C. and Foley, R. A. (1999). Species richness, species range size and ecological specialisation among African primates: geographical patterns and conservation implications. *Biodiversity and Conservation*, **8**, 1033–1056.

Eeley, H. A. C. and Lawes, M. J. (1999). Large-scale patterns of species richness and species range size in anthropoid primates. In *Primate communities* (eds J. G. Fleagle, C. Janson and K. E. Reed), pp. 191–219. Cambridge University Press, Cambridge.

Ehleringer, J. and House, D. (1984). Orientation and slope preference in barrel cactus (*Ferocactus acanthodes*) at its northern distribution limit. *Great Basin Naturalist*, **44**, 133–139.

Ehrlich, P. R. (1994). Energy use and biodiversity loss. *Philosophical Transactions of the Royal Society, London*, B **344**, 99–104.

Ehrlich, P. R. (1995). The scale of the human enterprise and biodiversity loss. In *Extinction rates* (eds J. H. Lawton and R. M. May), pp. 214–226. Oxford University Press, Oxford.

Ekbom, B. S. (1987). Incidence counts for estimating densities of *Rhopalosiphum padi* (Homoptera: Aphididae). *Journal of Economic Entomology*, **80**, 933–935.

Eldredge, N. (1999). Cretaceous meteor showers, the human ecological 'niche', and the sixth extinction. In *Extinctions in near time* (ed. R. D. E. MacPhee), pp. 1–15. Kluwer Academic, New York.

Elkins, N. (1995). *Weather and bird behaviour*, 2nd edn. Poyser, London.

Ellstrand, N. C., and Elam, D. R. (1993). Population genetic consequences of small population size: implications for plant conservation. *Annual Review of Ecology and Systematics*, **24**, 217–242.

Elmes, G. W., Wardlaw, J. C., Nielsen, M. G., Kipyatkov, V. E., Lopatina, E. B., Radchenko, A. G. and Barr, B. (1999). Site latitude influences on respiration rate, fat content and the ability of worker ants to rear larvae: a comparison of *Myrmica rubra* (Hymenoptera: Formicidae) populations over their European range. *European Journal of Entomology*, **96**, 117–124.

Elmhagen, B. and Angerbjörn, A. (2001). The applicability of metapopulation theory to large mammals. *Oikos*, **94**, 89–100.

Elton, C. (1927). *Animal ecology*. Sidgwick and Jackson, London.

Emlen, J. T., De Jong, M. J., Jaeger, M. J., Moermond, T. C., Rusterholz, K. A. and White, R. P. (1986). Density trends and range boundary constraints of forest birds along a latitudinal gradient. *Auk*, **103**, 791–803.

Emlet, R. B. (1995). Developmental mode and species geographic range in regular sea urchins (Echinodermata: Echinoidea). *Evolution*, **49**, 476–489.

Endler, J. A. (1977). *Geographic variation, speciation and gene flow*. Princeton University Press, Princeton, New Jersey.

Enquist, B. J., Jordan, M. A. and Brown, J. H. (1995). Connections between ecology, biogeography, and paleobiology: relationship between local abundance and geographic distribution in fossil and recent molluscs. *Evolutionary Ecology*, **9**, 586–604.

Erasmus, B. F. N., Kshatriya, M., Mansell, M. W., Chown, S. L. and van Jaarsveld, A. S. (2000). A modelling approach to antlion (Neuroptera: Myrmeleontidae) distribution patterns. *African Entomology*, **8**, 157–168.

Erickson, R. O. (1945). The *Clematis fremontii* var. *riehlii* population in the Ozarks. *Annals of Missouri Botanical Garden*, **32**, 413–460.

Eriksson, Å. (1998). Regional distribution of *Thymus serpyllum*: management history and dispersal limitation. *Ecography*, **21**, 35–43.

Erwin, D. H. (1996). Understanding biotic recoveries: extinction, survival and preservation during the End-Permian mass extinction. In *Evolutionary paleobiology* (eds D. Jablonski, D. H. Erwin and J. H. Lipps), pp. 398–418. University of Chicago Press, Chicago.

Etter, R. J., Rex, M. A., Chase, M. C. and Quattro, J. M. (1999). A genetic dimension to deep-sea biodiversity. *Deep-Sea Research I*, **46**, 1095–1099.

Eyre, M. D., Carr, R., McBlane, R. P. and Foster, G. N. (1992a). The effects of varying site-water duration on the distribution of water beetle assemblages, adults and larvae (Coleoptera: Haliplidae, Dytiscidae, Hydrophilidae). *Archiv für Hydrobiologie*, **124**, 281–291.

Eyre, M. D., Foster, G. N. and Young, A. G. (1993). Relationships between water-beetle distributions and climatic variables: a possible index for monitoring global climate change. *Archiv für Hydrobiologie*, **127**, 437–450.

Eyre, M. D., Rushton, S. P., Young, A. G. and Hill, D. (1992b). Land cover and breeding birds. In *Land use change: the causes and consequences* (ed. M. C. Whitby), pp. 131–136. HMSO, London.

Faith, D. P., Margules, C. R., Walker, P. A., Stein, J. and Natera, G. (2000a). Practical application of biodiversity surrogates and percentage targets for reservation in Papua New Guinea. *Pacific Conservation Biology*, **6**, 289–303.

Faith, D. P., Margules, C. R. and Walker, P. A. (2000b). A biodiversity conservation plan for Papua New Guinea based on biodiversity trade-offs analysis. *Pacific Conservation Biology*, **6**, 304–324.

Farlow, J. O. (1993). On the rareness of big, fierce animals: speculations about the body sizes, population densities, and geographic ranges of predatory mammals and large carnivorous dinosaurs. *American Journal of Science*, **293A**, 167–199.

FAUNMAP Working Group (Graham, R. W., Lundelius, E. L. Jr, Graham, M. A., Schroeder, E. K., Toomey, R. S. III, Anderson, E., Barnosky, A. D., Burns, J. A., Churcher, C. S., Grayson, D. K., Guthrie, R. D., Harington, C. R., Jefferson, G. T., Martin, L. D., McDonald, H. G., Morlan, R. E., Semken, H. A. Jr, Webb, S. D., Werdelin, L. and Wilson, M. C. (1996). Spatial response of mammals to Late Quaternary environmental fluctuations. *Science*, **272**, 1601–1606.

Fenchel, T. (1993). There are more small then large species? *Oikos*, **68**, 375–378.

Feng, M. C., Nowierski, R. M. and Zeng, Z. (1993). Populations of *Sitobion avenae* and *Aphidius ervi* on spring wheat in the northwestern United States. *Entomologia Experimentalis et Applicata*, **67**, 109–117.

Fernández-Juricic, E. (2000). Bird community composition patterns in urban parks of Madrid: the role of age, size and isolation. *Ecological Research*, **15**, 373–383.

Ferrer, X., Motis, A. and Peris, S. J. (1991). Changes in the breeding range of starlings in the Iberian Peninsula during the last 30 years: competition as a limiting factor. *Journal of Biogeography*, **18**, 631–636.

Finlay, B. J. and Clarke, K. J. (1999). Ubiquitous dispersal of microbial species. *Nature*, **400**, 828.

Finlay, B. J., Esteban, G. F. and Fenchel, T. (1996). Global diversity and body size. *Nature*, **383**, 132–133.

Finlay, B. J., Esteban, G. F., Olmo, J. L. and Tyler, P. A. (1999). Global distribution of free-living microbial species. *Ecography*, **22**, 138–144.

Finlay, B. J. and Fenchel, T. (1999). Divergent perspectives on protist species richness. *Protist*, **150**, 229–233.

Finlayson, J. C. (1999). Species abundances across spatial scales. *Science*, **283**, 1979.

Fischer, J. and Lindenmayer, D. B. (2000). An assessment of the published results of animal relocations. *Biological Conservation*, **96**, 1–11.

Fischer, M. and Matthies, D. (1998). Effects of population size on performance in the rare plant *Gentianella germanica*. *Journal of Ecology*, **86**, 195–204.

Fitt, G. P. (1989). The ecology of *Heliothis* species in relations to agroecosystems. *Annual Review of Entomology*, **34**, 17–52.

Flannery, T. (2001). *The eternal frontier: an ecological history of North America and its peoples*. Heinemann, London.

Flebbe, P. A. (1993). Comment on Meisner (1990): effect of climatic warming on the southern margins of the native range of brook trout, *Salvelinus fontinalis*. *Canadian Journal of Fisheries and Aquatic Science*, **50**, 883–884.

Flebbe, P. A. (1994). A regional view of the margin: salmonid abundance and distribution in the southern Appalachian Mountains of North Carolina and Virginia. *Transactions of the American Fisheries Society*, **123**, 657–667.

Fleming, I. A. and Gross, M. R. (1990). Latitudinal clines: a trade-off between egg number and size in Pacific Salmon. *Ecology*, **71**, 1–11.

Flessa, K. W. and Jablonski, D. (1996). The geography of evolutionary turnover: a global analysis of extant bivalves. In *Evolutionary paleobiology* (eds D. Jablonski, D. H. Erwin and J. H. Lipps), pp. 376–397. University of Chicago Press, Chicago.

Flessa, K. W. and Thomas, R. H. (1985). Modelling the biogeographic regulation of evolutionary rates. In *Phanerozoic diversity patterns. Profiles in macroevolution* (ed. J. W. Valentine), pp. 355–376. Princeton University Press, Princeton.

Forcella, F. and Wood, J. T. (1984). Colonization potentials of alien weeds are related to their 'native' distributions: implications for plant quarantine. *Journal of the Australian Institute of Agricultural Science*, **50**, 35–41.

Forcella, F., Wood, J. T. and Dillon, S. P. (1986). Characteristics distinguishing invasive weeds within *Echium* (Bugloss). *Weed Research*, **26**, 351–364.

Ford, H. A. (1990). Relationships between distribution, abundance and foraging specialization in Australian landbirds. *Ornis Scandinavica*, **21**, 133–138.

Ford, M. J. (1982). *The changing climate: responses of the natural fauna and flora*. George Allen and Unwin, London.

Fox, L. R. and Morrow, P. A. (1981). Specialisation: species property or local phenomenon? *Science*, **211**, 887–893.

Fox, R., Warren, M., Harding, P., Asher, J., Jeffcoate, G. and Jeffcoate, S. (2001). The millennium atlas of butterflies in Britain and Ireland. *British Wildlife*, **12**, 173–178.

France, R. (1992). The North American latitudinal gradient in species richness and geographical range of freshwater crayfish and amphipods. *American Naturalist*, **139**, 342–354.

Frankham, R. (1996). Relationship of genetic variation to population size in wildlife. *Conservation Biology*, **10**, 1500–1508.

Freitag, R. (1969). A revision of the species of the genus *Evarthrus* LeConte (Coleoptera: Carabidae). *Quaestiones Entomologicae*, **5**, 89–212.

Fretwell, S. D. and Lucas, H. L. (1970). On territorial behaviour and other factors influencing habitat distribution in birds. *Acta Biotheoriologica*, **19**, 16–36.

Frey, J. K. (1992). Response of a mammalian faunal element to climatic changes. *Journal of Mammalogy*, **73**, 43–50.

Frey, J. K. (1993). Modes of peripheral isolate formation and speciation. *Systematic Biology*, **42**, 373–381.

Fukuda, I. and Grant, W. F. (1980). Chromosome variation and evolution in *Trillium grandiflorum*. *Canadian Journal of Genetics and Cytology*, **22**, 81–91.

Fuller, R. J. (1982). *Bird habitats in Britain*. Poyser, Calton, Staffordshire.

Fuller, R. J., Gregory, R. D., Gibbons, D. W., Marchant, J. H., Wilson, J. D., Baillie, S. R. and Carter, N. (1995). Population declines and range contractions among lowland farmland birds in Britain. *Conservation Biology*, **9**, 1425–1441.

Furnier, G. R. and Adams, W. T. (1986). Geographic patterns of allozyme variation in Jeffrey pine. *American Journal of Botany*, **73**, 1009–1015.

García, D., Zamora, R., Gómez, J. M., Jordano, P. and Hódar, J. A. (2000). Geographical variation in seed production, predation and abortion in *Juniperus communis* throughout its range in Europe. *Journal of Ecology*, **88**, 436–446.

García-Ramos, G. and Kirkpatrick, M. (1997). Genetic models of adaptation and gene flow in peripheral populations. *Evolution*, **51**, 21–28.

García-Ramos, G., Sáchez-Garduño, F. and Maini, P. K. (2000). Dispersal can sharpen parapatric boundaries on a spatially varying environment. *Ecology*, **81**, 749–760.

Garland, S. P. (1982). *Butterflies of the Sheffield area*. Sorby Natural History Society, Sheffield.

Garrett, W. E. (ed.) (1988). *Historical atlas of the United States*. National Geographic Society, Washington D. C.

Gasc, J.-P., Cabela, A., Crnobrnja-Isailovic, J., Dolmen, D., Grossenbacher, K., Haffner, P., Lescure, J., Martens, H., Martínez Rica, J. P., Maurin, H., Oliveira, M. E., Sofianidou, T. S., Veith, M. and Zuiderwijk, A. (eds) (1997). *Atlas of amphibians and reptiles in Europe*. Societas Europaea Herpetologica and Muséum National d'Histoire Naturelle (IEGB/SPN), Paris.

Gaston, K. J. (1990). Patterns in the geographical ranges of species. *Biological Reviews*, **65**, 105–129.

Gaston, K. J. (1991a). How large is a species' geographic range? *Oikos*, **61**, 434–438.

Gaston, K. J. (1991b). The magnitude of global insect species richness. *Conservation Biology*, **5**, 283–296.

Gaston, K. J. (1994a). *Rarity*. Chapman and Hall, London.

Gaston, K. J. (1994b). Measuring geographic range sizes. *Ecography*, **17**, 198–205.

Gaston, K. J. (1994c). Geographic range sizes and trajectories to extinction. *Biodiversity Letters*, **2**, 163–170.

Gaston, K. J. (ed) (1996a). *Biodiversity: a biology of numbers and difference*. Blackwell Science, Oxford.

Gaston, K. J. (1996b). Species richness: measure and measurement. In *Biodiversity: a biology of numbers and difference* (ed. K. J. Gaston), pp. 77–113. Blackwell Science, Oxford.

Gaston, K. J. (1996c). Species-range size distributions: patterns, mechanisms and implications. *Trends in Ecology and Evolution*, **11**, 197–201.

Gaston, K. J. (1996d). The multiple forms of the interspecific abundance-distribution relationship. *Oikos*, **75**, 211–220.

Gaston, K. J. (1998). Species-range size distributions: products of speciation, extinction and transformation. *Philosophical Transactions of the Royal Society, London*, **353**, 219–230.

Gaston, K. J. and Blackburn, T. M. (1996a). Range size-body size relationships: evidence of scale dependence. *Oikos*, **75**, 479–485.

Gaston, K. J. and Blackburn, T. M. (1996b). Global scale macroecology: interactions between population size, geographic range size and body size in the Anseriformes. *Journal of Animal Ecology*, **65**, 701–714.

Gaston, K. J. and Blackburn, T. M. (1996c). Conservation implications of geographic range size-body size relationships. *Conservation Biology*, **10**, 638–646.

Gaston, K. J. and Blackburn, T. M. (1997a). Evolutionary age and risk of extinction: the global avifauna. *Evolutionary Ecology*, **11**, 557–565.

Gaston, K. J. and Blackburn, T. M. (1997b). Age, area and avian diversification. *Biological Journal of the Linnean Society*, **62**, 239–253.

Gaston, K. J. and Blackburn, T. M. (2000). *Pattern and process in macroecology*. Blackwell Science, Oxford.

Gaston, K. J., Blackburn, T. M., Greenwood, J. J. D., Gregory, R. D., Quinn, R. M. and Lawton, J. H. (2000). Abundance-occupancy relationships. *Journal of Applied Ecology*, **37** (Suppl. 1), 39–59.

Gaston, K. J., Blackburn, T. M. and Gregory, R. D. (1997b). Abundance-range size relationships of breeding and wintering birds in Britain: a comparative analysis. *Ecography*, **20**, 569–579.

Gaston, K. J., Blackburn, T. M. and Gregory, R. D. (1997c). Interspecific abundance-range size relationships: range position and phylogeny. *Ecography*, **20**, 390–399.

Gaston, K. J., Blackburn, T. M. and Gregory, R. D. (1998d). Interspecific differences in intraspecific abundance-range size relationships of British breeding birds. *Ecography*, **21**, 149–158.

Gaston, K. J., Blackburn, T. M., Gregory, R. D. and Greenwood, J. J. D. (1998e). The anatomy of the interspecific abundance-range size relationship for the British avifauna: I. Spatial patterns. *Ecology Letters*, **1**, 38–46.

Gaston, K. J., Blackburn, T. M. and Gregory, R. D. (1999a). Intraspecific abundance-occupancy relationships: case studies of six bird species in Britain. *Diversity and Distributions*, **5**, 197–212.

Gaston, K. J., Blackburn, T. M. and Lawton, J. H. (1997a). Interspecific abundance-range size relationships: an appraisal of mechanisms. *Journal of Animal Ecology*, **66**, 579–601.

Gaston, K. J., Blackburn, T. M. and Lawton, J. H. (1998c). Aggregation and interspecific abundance-occupancy relationships. *Journal of Animal Ecology*, **67**, 995–999.

Gaston, K. J., Blackburn, T. M. and Spicer, J. I. (1998b). Rapoport's rule: time for an epitaph? *Trends in Ecology and Evolution*, **13**, 70–74.

Gaston, K. J. and Chown, S. L. (1999a). Geographic range size and speciation. In *Evolution of biological diversity* (eds A. E. Magurran and R. M. May), pp. 236–259. Oxford University Press, Oxford.

Gaston, K. J. and Chown, S. L. (1999b). Why Rapoport's rule does not generalise. *Oikos*, **84**, 309–312.

Gaston, K. J. and Curnutt, J. L. (1998). The dynamics of abundance-range size relationships. *Oikos*, **81**, 38–44.

Gaston, K. J., Gregory, R. D. and Blackburn, T. M. (1999b). Intraspecific relationships between abundance and occupancy among species of Paridae and Sylviidae in Britain. *Écoscience*, **6**, 131–142.

Gaston, K. J. and He, F. (2002). The distribution of species range size: a stochastic process. *Proceedings of the Royal Society, London B*, **269**, 1079–1086.

Gaston, K. J. and Kunin, W. E. (1997). Concluding comments. In *The biology of rarity: causes and consequences of rare-common differences* (eds W. E. Kunin and K. J. Gaston), pp. 262–272. Chapman and Hall, London.

Gaston, K. J. and Lawton, J. H. (1988). Patterns in body size, population dynamics and regional distribution of bracken herbivores. *American Naturalist*, **132**, 662–680.

Gaston, K. J. and Lawton, J. H. (1990). Effects of scale and habitat on the relationship between regional distribution and local abundance. *Oikos*, **58**, 329–335.

Gaston, K. J. and McArdle, B. H. (1994). The temporal variability of animal abundances: measures, methods and patterns. *Philosophical Transactions of the Royal Society, London*, B **345**, 335–358.

Gaston, K. J., Quinn, R. M., Blackburn, T. M. and Eversham, B. C. (1988a). Species–range size distributions in Britain. *Ecography*, **21**, 361–370.

Gaston, K. J., Quinn, R. M., Wood, S. and Arnold, H. R. (1996). Measures of geographic range size: the effects of sample size. *Ecography*, **19**, 259–268.

Gaston, K. J. and Spicer, J. I. (1998). *Biodiversity: an introduction*. Blackwell Science, Oxford.

Gaston, K. J. and Rodrigues, A. S. L. (2003). Reserve selection in regions with poor biological data. *Conservation Biology*, in press.

Gates, S. and Donald, P. F. (2000). Local extinction of British farmland birds and the prediction of further loss. *Journal of Applied Ecology*, **37**, 806–820.

Gauld, I. D. and Gaston, K. J. (1995). The Costa Rican hymenopteran fauna. In *The Hymenoptera of Costa Rica*. (eds P. E. Hanson and I. D. Gauld), pp. 13–19. Oxford University Press, Oxford.

Gauld, I. D. and Mitchell, P. A. (1981). *The taxonomy, distribution and host preferences of Indo-Papuan parasitic wasps of the subfamily Ophioninae (Hymenoptera: Ichneumonidae)*. Commonwealth Agricultural Bureaux, Slough.

Gavrilets, S., Li, H. and Vose, M. D. (1998). Rapid parapatric speciation on holey adaptive landscapes. *Proceedings of the Royal Society, London*, B **265**, 1483–1489.

Gavrilets, S., Li, H. and Vose, M. D. (2000). Patterns of parapatric speciation. *Evolution*, **54**, 1126–1134.

Gaylord, B. and Gaines, S. D. (2000). Temperature or transport? Range limits in marine species mediated solely by flow. *American Naturalist*, **155**, 769–789.

Gear, A. J. and Huntley, B. (1991). Rapid changes in the range limits of Scots pine 4000 years ago. *Science*, **251**, 544–547.

George, V. S. (1985). Demographic evaluation of the influence of temperature and salinity on the copepod *Eurytemora herdmani*. *Marine Ecology Progress Series*, **21**, 145–152.

Gibbons, D. W., Reid, J. B. and Chapman, R. A. (1993). *The new atlas of breeding birds in Britain and Ireland: 1988–1991*. Poyser, London.

Gibson, R. N. (1994). Impact of habitat quality and quantity on the recruitment of juvenile flatfishes. *Netherlands Journal of Sea Research*, **32**, 191–206.

Gilbert, N. (1980). Comparative dynamics of a single-host aphid. I. The evidence. *Journal of Animal Ecology*, **49**, 351–369.

Gillespie, T. W. (2000). Rarity and conservation of forest birds in the tropical dry forest region of Central America. *Biological Conservation*, **96**, 161–168.

Gillis, D. M., Kramer, D. L. and Bell, G. (1986). Taylor's power law as a consequence of Fretwell's Ideal Free Distribution. *Journal of Theoretical Biology*, **123**, 281–287.

Gioia, P. and Pigott, J. P. (2000). Biodiversity assessment: a case study in predicting richness from the potential distributions of plant species in the forests of south-western Australia. *Journal of Biogeography*, **27**, 1065–1078.

Glazier, D. S. (1980). Ecological shifts and the evolution of geographically restricted species of North American *Peromyscus* (mice). *Journal of Biogeography*, **7**, 63–83.

Glazier, D. S. (1987). Toward a predictive theory of speciation—the ecology of isolate selection. *Journal of Theoretical Biology*, **126**, 323–333.

Gleason, H. A. (1924). Age and area from the viewpoint of phytogeography. *American Journal of Botany*, **11**, 541–546.

Godt, M. J. W., Hamrick, J. L. and Bratton, S. (1995). Genetic diversity in a threatened wetland species, *Helonias bullata* (Liliaceae). *Conservation Biology*, **9**, 596–604.

Gofas, S. (1998). Marine molluscs with a very restricted range in the Strait of Gibraltar. *Diversity and Distributions*, **4**, 255–266.

Gómez, J. M. (1996). Predispersal reproductive ecology of an arid land crucifer, *Moricandia moricandioides*: effect of mammal herbivory on seed production. *Journal of Arid Environments*, **33**, 425–437.

Gonzalez, A., Lawton, J. H., Gilbert, F. S., Blackburn, T. M. and Evans-Freke, I. (1998). Metapopulation dynamics maintain the positive species abundance-distribution relationship. *Science*, **281**, 2045–2047.

Good, R. D'O. (1931). A theory of plant geography. *New Phytologist*, **30**, 149–171.

Gooday, A. J. (1999). Biodiversity of Foraminifera and other protests in the deep sea: scales and patterns. *Belgium Journal of Zoology*, **129**, 61–80.

Gooday, A. J., Bett, B. J., Shires, R. and Lambshead, P. J. D. (1998). Deep-sea benthic foraminiferal species diversity in the NE Atlantic and NW Arabian sea: a synthesis. *Deep-Sea Research II*, **45**, 165–201.

Goodbody, I. (1961). Mass mortality of a marine fauna following tropical rains. *Ecology*, **42**, 150–155.

Goodman, D. (1987). The demography of chance extinction. In *Viable populations for conservation* (ed. M. E. Soulé), pp. 11–34. Cambridge University Press, Cambridge.

Gorodkov, K. B. (1986). Three-dimensional climatic model of potential range and some of its characteristics. II. *Entomological Review*, **65**, 1–18.

Gotelli, N. J. (2001). Research frontiers in null model analysis. *Global Ecology and Biogeography*, **10**, 337–343.

Gotelli, N. J. and Simberloff, D. (1987). The distribution and abundance of tallgrass prairie plants: A test of the core-satellite hypothesis. *American Naturalist*, **130**, 18–35.

Grace, J. (1987). Climatic tolerance and the distribution of plants. *New Phytologist*, **106** (Suppl.), 113–130.

Gram, W. K. and Sork, V. L. (1999). Population density as a predictor of genetic variation for woody plant species. *Conservation Biology*, **13**, 1079–1087.

Gray, R. D. and Craig, J. L. (1991). Theory really matters: hidden assumptions in the concept of 'habitat requirements'. *Acta XX Congressus Internationalis Ornithologici*, 2553–2560.

Grayson, D. K. (2001). The archaeological record of human impacts on animal populations. *Journal of World Prehistory*, **15**, 1–68.

Green, D. M., Sharbel, T. F., Kearsley, J. and Kaiser, H. (1996). Postglacial range fluctuation, genetic subdivision and speciation in the western North American spotted frog complex, *Rana pretiosa*. *Evolution*, **50**, 374–390.

Green, R. E. (1996). Factors affecting the population density of the corncrake *Crex crex* in Britain and Ireland. *Journal of Applied Ecology*, **33**, 237–248.

Gregory, R. D. (1990). Parasites and host geographic range as illustrated by waterfowl. *Functional Ecology*, **4**, 645–654.

Gregory, R. D. (1994). Species abundance patterns of British birds. *Proceedings of the Royal Society, London*, B **257**, 299–301.

Gregory, R. D. (1995). Phylogeny and relations among abundance, geographical range and body size of British breeding birds. *Philosophical Transactions of the Royal Society, London*, B **349**, 345–351.

Gregory, R. D., Bashford, R. I. and Balmer, D. B. *et al.* (1997). *The breeding bird survey (1995–1996)*. British Trust for Ornithology, Thetford.

Gregory, R. D. and Blackburn, T. M. (1998). Macroecological patterns in British breeding birds: covariation of species' geographical range sizes at differing spatial scales. *Ecography*, **21**, 527–534.

Gregory, R. D. and Gaston, K. J. (2000). Explanations of commonness and rarity in British breeding birds: separating resource use and resource availability. *Oikos*, **88**, 515–526.

Gregory, R. D., Greenwood, J. J. D. and Hagemeijer, E. J. M. (1998). The EBCC atlas of European breeding birds: a contribution to science and conservation. *Biologia e Conservazione della Fauna*, **102**, 38–49.

Greuter, W. (1991). Botanical diversity, endemism, rarity, and extinction in the Mediterranean area: an analysis based on the published volumes of Med-checklist. *Botanist's Chronicle*, **10**, 63–79.

Griggs, R. F. (1914). Observations on the behavior of some species at the edges of their ranges. *Bulletin of the Torrey Botanical Club*, **41**, 25–49.

Grime, J. P., Hodgson, J. G. and Hunt, R. (1988). *Comparative plant ecology: a functional approach to common British species*. Unwin Hyman, London.

Grosholz, E. D. (1996). Contrasting rates of spread for introduced species in terrestrial and marine systems. *Ecology*, **77**, 1680–1686.

Gross, S. J. and Price, T. D. (2000). Determinants of the northern and southern range limits of a warbler. *Journal of Biogeography*, **27**, 869–878.

Grubb, P. J. (1989). Toward a more exact ecology: a personal view of the issues. In *Toward a more exact ecology* (eds P. J. Grubb and J. B. Whittaker), pp. 3–29. Blackwell Scientific, Oxford.

Guries, R. P. and Ledig, F. T. (1982). Genetic diversity and population-structure in pitch pine (*Pinus rigida* Mill). *Evolution*, **36**, 387–402.

Gutiérrez, D. and Menéndez, R. (1997). Patterns in the distribution, abundance and body size of carabid beetles (Coleoptera: Caraboidea) in relation to dispersal ability. *Journal of Biogeography*, **24**, 903–914.

Gutiérrez, D. and Thomas, C. D. (2000). Marginal range expansion in a host-limited butterfly species *Gonepteryx rhamni*. *Ecological Entomology*, **25**, 165–170.

Gyllenberg, M. and Hanski, I. (1992). Single-species metapopulation dynamics: a structured model. *Theoretical Population Biology*, **42**, 35–66.

Haapala, J. and Saurola, P. (1995). Constant Effort Sites Scheme in Finland 1992–94. *Linnut*, **3**, 32–33.

Haftorn, S. (1978). Energetics of incubation by the Goldcrest *Regulus regulus* in relation to ambient air temperatures and the geographical distribution of the species. *Ornis Scandinavica*, **9**, 22–30.

Hagemeijer, E. J. M. and Blair, M. J. (eds) (1997). *The EBCC atlas of European breeding birds: their distribution and abundance*. Poyser, London.

Haldane, J. B. S. (1956). The relation between density regulation and natural selection. *Proceedings of the Royal Society of London B*, **145**, 306–308.

Hall, C. A. S., Stanford, J. A. and Hauer, F. R. (1992). The distribution and abundance of organisms as a consequence of energy balances along multiple environmental gradients. *Oikos*, **65**, 377–390.

Hallingbäck, T. (1998). The new IUCN threat categories tested on Swedish bryophytes. *Lindbergia*, **23**, 13–27.

Hallingbäck, T., Hodgetts, N., Raeymaekers, G., Schumacker, R., Sérgio, C., Söderström, L., Stewart, N. and Váňa, J. (1998). Guidelines for application of the revised IUCN threat categories to bryophytes. *Lindbergia*, **23**, 6–12.

Halloy, S. R. P. (1999). The dynamic contribution of new crops to the agricultural economy: is it predictable? In *Perspectives on new crops and new uses* (ed. J. Janick.), pp. 53–59. ASHS Press, Alexandria, Virginia.

Hamrick, J. L., Blanton, H. M. and Hamrick, K. J. (1989). Genetic structure of geographically marginal populations of Ponderosa pine. *American Journal of Botany*, **76**, 1559–1568.

Hannah, L., Carr, J. L. and Lankerani, A. (1995). Human disturbance and natural habitat: a biome level analysis of a global data set. *Biodiversity and Conservation*, **4**, 128–155.

Hansen, P. A. (1989). Species response curves of macrofungi along a mull/mor gradient in Swedish beech forests. *Vegetatio*, **82**, 69–78.

Hansen, T. A. (1978). Larval dispersal and species longevity in Lower Tertiary gastropods. *Science*, **199**, 886–887.

Hansen, T. A. (1980). Influence of larval dispersal and geographic distribution on species longevities in neogastropods. *Paleobiology*, **6**, 193–207.

Hanski, I. (1982). Dynamics of regional distribution: the core and satellite species hypothesis. *Oikos*, **38**, 210–221.

Hanski, I. (1991a). Single-species metapopulation dynamics: concepts, models and observations. *Biological Journal of the Linnean Society*, **42**, 17–38.

Hanski, I. (1991b). Reply to Nee, Gregory and May. *Oikos*, **62**, 88–89.

Hanski, I. (1992). Inferences from ecological incidence functions. *American Naturalist*, **139**, 657–662.

Hanski, I. and Cambefort, Y. (1991). Spatial processes. In *Dung beetle ecology* (eds I. Hanski and Y. Cambefort), pp. 283–304. Princeton University Press, Princeton, New Jersey.

Hanski, I. and Gilpin, M. E. (eds) (1997). *Metapopulation biology: ecology, genetics, and evolution*. Academic Press, San Diego.

Hanski, I. and Gyllenberg, M. (1993). Two general metapopulation models and the core-satellite species hypothesis. *American Naturalist*, **142**, 17–41.

Hanski, I. and Gyllenberg, M. (1997). Uniting two general patterns in the distribution of species. *Science*, **275**, 397–400.

Hanski, I., Kouki, J. and Halkka, A. (1993). Three explanations of the positive relationship between distribution and abundance of species. In *Species diversity in ecological communities: historical and geographical perspectives* (eds R. E. Ricklefs and D. Schluter), pp. 108–116. University of Chicago Press, Chicago.

Harding, B. D. (1979). *Bedfordshire bird atlas*. Bedfordshire Natural History Society, Kettering.

Hardy, P. B. (1998). *Butterflies of Greater Manchester*. PGL Enterprises, Sale.

Harrison, J. A., Allan, D. G., Underhill, L. G., Herremans, M., Tree, A. J., Parker, V. and Brown, C. J. (eds) (1997a). *The atlas of southern African birds. Volume 1: Non-passerines*. BirdLife South Africa, Johannesburg.

Harrison, J. A., Allan, D. G., Underhill, L. G., Herremans, M., Tree, A. J., Parker, V. and Brown, C. J. (eds) (1997b). *The atlas of southern African birds. Volume 2: Passerines*. BirdLife South Africa, Johannesburg.

Harrison, R. D. (2000). Repercussions of El Niño: drought causes extinction and the breakdown of mutualism in Borneo. *Proceedings of the Royal Society, London B*, **267**, 911–915.

Harte, J., Blackburn, T. and Ostling, A. (2001). Self-similarity and the relationship between abundance and range size. *American Naturalist*, **157**, 374–386.

Harte, J., Kinzig, A. and Green, J. (1999). Self-similarity in the distribution and abundance of species. *Science*, **284**, 334–336.

Hartley, S. (1998). A positive relationship between local abundance and regional occupancy is almost inevitable (but not all positive relationships are the same). *Journal of Animal Ecology*, **67**, 992–994.

Harvey, P. H. (1996). Phylogenies for ecologists. *Journal of Animal Ecology*, **65**, 255–263.

Harvey, P. H. and Pagel, M. D. (1991). *The comparative method in evolutionary biology*. Oxford University Press, Oxford.

Hassell, M. P., Southwood, T. R. E. and Reader, P. M. (1987). The dynamics of the viburnum whitefly (*Aleurotrachelus jelinekii*): a case study of population regulation. *Journal of Animal Ecology*, **56**, 283–300.

Hausdorf, B. (2000). Biogeography of the Limacoidea *sensu lato* (Gastropoda: Styommatophora): vicariance events and long-distance dispersal. *Journal of Biogeography*, **27**, 379–390.

Hawkins, J. P., Roberts, C. M. and Clark, V. (2000). The threatened status of restricted-range coral reef fish species. *Animal Conservation*, **3**, 81–88.

Hayden, B. P. (1988). Ecosystem feedbacks on climate at the landscape scale. *Philosophical Transactions of the Royal Society, London B*, **353**, 5–18.

He, F. and Gaston, K. J. (2000). Occupancy-abundance relationships and sampling scales. *Ecography*, **23**, 503–511.

He, F., Gaston, K. J. and Wu, J. (2002). On species occupancy-abundance models. *Écoscience* **9**, 119–126.

Heath, J., Pollard, E. and Thomas, J. A. (1984). *Atlas of butterflies in Britain and Ireland*. Penguin, London.

Hecnar, S. J. (1999). Patterns of turtle species' geographic range size and a test of Rapoport's rule. *Ecography*, **22**, 436–446.

Hector, A., Schmid, B., Beierkuhnlein, C., Caldeira, M. C., Diemer, M., Dimitrakopoulos, P. G., Finn, J., Freitas, H., Giller, P. S., Good, J., Harris, R., Högberg, P., Huss-Danell, K., Joshi, J., Jumpponen, A., Körner, C., Leadley, P. W., Loreau, M., Minns, A., Mulder, C. P. H., O'Donovan, G., Otway, S. J., Pereira, J. S., Prinz, A., Read, D. J., Scherer-Lorenzen, M., Schulze, E.-D., Siamantziouras, A-S. D., Spehn, E. M., Terry, A. C., Troumbis, A. Y., Woodward, F. I., Yachi, S. and Lawton, J. H. (1999). Plant diversity and productivity experiments in European grasslands. *Science*, **286**, 1123–1127.

Heikkinen, J. and Högmander, H. (1994). Fully Bayesian-approach to image restoration with an application in biogeography. *Applied Statistics–Journal of Royal Statistical Society*, C **43**, 569–582.

Henderson, P. A. and Seaby, R. M. (1999). Population stability of the sea snail at the southern edge of its range. *Journal of Fish Biology*, **54**, 1161–1176.

Hendricks, P. (1997). Geographical trends in clutch size: a range-wide relationship with laying date in American pipits. *Auk*, **114**, 773–778.

Hengeveld, R. (1989). *Dynamics of biological invasions*. Chapman and Hall, London.

Hengeveld, R. (1990a). *Dynamic biogeography*. Cambridge University Press, Cambridge.

Hengeveld, R. (1990b). Theories on species responses to variable climates. In *Landscape-ecological impact of climate changes* (eds M. M. Boer and R. S. de Groot), pp. 274–289. IOS Press, Amsterdam.

Hengeveld, R. and Haeck, J. (1982). The distribution of abundance. I. Measurements. *Journal of Biogeography*, **9**, 303–316.

Hengeveld, R. and Hogeweg, P. (1979). Cluster analysis of the distribution patterns of Dutch carabid species (Col.). In *Multivariate methods in ecological work* (eds L. Orloci, C. R. Rao and W. M. Stiteler), pp. 65–86. International Co-operative Publishing House, Fairland, MD.

Hengeveld, R., Kooijman, S. A. L. M. and Taillie, C. (1979). A spatial model explaining species-abundance curves. In *Statistical distributions in ecological work* (eds J. K. Ord, G. P. Patil and C. Taillie), pp. 333–347. International Co-operative Publishing House, Fairland, Maryland.

Hepworth, G. and MacFarlane, J. R. (1992). Systematic presence-absence sampling method applied to two-spotted spider mite (Acari: Tetranychidae) on strawberries in Victoria, Australia. *Journal of Economic Entomology*, **85**, 2234–2239.

Hersteinsson, P. and Macdonald, D. W. (1992). Interspecific competition and the geographical distribution of red and arctic foxes *Vulpes vulpes* and *Alopex lagopus*. *Oikos*, **64**, 505–515.

Hess, G. K., West, R. L., Barnhill III, M. V. and Fleming, L. M. (2000). *Birds of Delaware*. University of Pittsburgh Press, Pittsburgh.

Hesse, R., Allee, W. C. and Schmidt, K. P. (1937). *Ecological animal geography*. Wiley, New York.

Hewitt, G. (2000). The genetic legacy of the Quaternary ice ages. *Nature*, **405**, 907–913.

Heywood, V. H. (ed.) (1995). *Global biodiversity assessment*. Cambridge University Press, Cambridge.

Higashi, M., Takimoto, G. and Yamamura, N. (1999). Sympatric speciation by sexual selection. *Nature*, **402**, 523–526.

Higgins, S. I., Richardson, D. M., Cowling, R. M. and Trinder-Smith, T. H. (1999). Predicting the landscape-scale distribution of alien plants and their threat to plant diversity. *Conservation Biology*, **13**, 303–313.

Higuchi, H. (1986). Bait-fishing by the Green-backed Heron *Ardeola striata* in Japan. *Ibis*, **128**, 285–290.

Hill, J. K., Thomas, C. D. and Huntley, B. (1999). Climate and habitat availability determine 20th century changes in a butterfly's range margin. *Proceedings of the Royal Society, London B*, **266**, 1197–1206.

Hilton-Taylor, C. (2000). *The (2000) IUCN red list of threatened species*. IUCN, Gland.

Hinsley, S. A., Pakeman, R., Bellamy, P. E. and Newton, I. (1996). Influences of habitat fragmentation on bird species distributions and regional population size. *Proceedings of the Royal Society, London B*, **263**, 307–313.

Hitchens, S. P. and Beebee, T. J. C. (1996). Persistence of British natterjack toad *Bufo calamita* Laurenti (Anura: Bufonidae) populations despite low genetic diversity. *Biological Journal of the Linnean Society*, **57**, 69–80.

Hobbs, R. and Lleras, E. (1995). Protecting and restoring ecosystems, species, populations and genetic diversity. In *Global biodiversity assessment* (eds V. H. Heywood and R. T. Watson), pp. 981–1017. UNEP, Cambridge University Press.

Hoch, H. and Howarth, F. G. (1989). 6 new cavernicolous cixiid planthoppers in the genus *Solonaima* from Australia (Homoptera, Fulgoroidea). *Systematic Entomology*, **14**, 377–402.

Hochberg, M. E. and Ives, A. R. (1999). Can natural enemies enforce geographical range limits? *Ecography*, **22**, 268–276.

Hocker, H. W. Jr (1956). Certain aspects of climate as related to the distribution of loblolly pine. *Ecology*, **37**, 824–834.

Hodgson, J. G. (1986). Commonness and rarity in plants with special reference to the Sheffield flora. Part II: The relative importance of climate, soils and land use. *Biological Conservation*, **36**, 253–274.

Hodgson, J. G. (1991). Management for the conservation of plants with particular reference to the British flora. In *The scientific management of temperate communities for conservation* (eds I. F. Spellerberg, F. B. Goldsmith and M. G. Morris), pp. 81–102. Blackwell Scientific, Oxford.

Hodgson, J. G. (1993). Commonness and rarity in British butterflies. *Journal of Applied Ecology*, **30**, 407–427.

Hodkinson, D. J. and Thompson, K. (1997). Plant dispersal: the role of man. *Journal of Applied Ecology*, **34**, 1484–1496.

Hodkinson, I. D. (1997). Progressive restriction of host plant exploitation along a climatic gradient: the willow psyllid *Cacopsylla groenlandica* in Greenland. *Ecological Entomology*, **22**, 47–54.

Hodkinson, I. D. (1999). Species response to global environmental change or why ecophysiological models are important: a reply to Davis *et al. Journal of Animal Ecology*, **68**, 1259–1262.

Hoffmann, A. A. and Blows, M. W. (1994). Species borders: ecological and evolutionary perspectives. *Trends in Ecology and Evolution*, **9**, 223–227.

Hoffmann, A. A. and Parsons, P. A. (1997). *Extreme environmental change and evolution.* Cambridge University Press, Cambridge.

Högmander, H. and Møller, J. (1995). Estimating distribution maps from atlas data using methods of statistical image analysis. *Biometrics*, **51**, 393–404.

Holm, L. G., Plucknett, D. L., Pancho, J. V. and Herberger, J. P. (1977). *The World's worst weeds: distribution and biology.* University Press of Hawaii, Honolulu.

Holt, R. D. (1977). Predation, apparent competition, and the structure of prey communities. *Theoretical Population Biology*, **12**, 197–229.

Holt, R. D. and Keitt, T. H. (2000). Alternative causes for range limits: a metapopulation perspective. *Ecology Letters*, **3**, 41–47.

Holt, R. D. and Lawton, J. H. (1994). The ecological consequences of shared natural enemies. *Annual Review of Ecology and Systematics*, **25**, 495–520.

Holt, R. D., Lawton, J. H., Gaston, K. J. and Blackburn, T. M. (1997). On the relationship between range size and local abundance: back to basics. *Oikos*, **78**, 183–190.

Horwood, J. W. and Millner, R. S. (1998). Cold induced abnormal catches of sole. *Journal of the Marine Biological Association, U. K.*, **78**, 345–347.

Houle, G. and Bouchard, F. (1990). Hackberry (*Celtis occidentalis*) at the northeastern limit of its distribution in North America: population structure and radial growth patterns. *Canadian Journal of Botany*, **68**, 2685–2692.

Houle, G. and Filion, L. (1993). Interannual variations in the seed production of *Pinus banksiana* at the limit of the species distribution in northern Québec, Canada. *American Journal of Botany*, **80**, 1242–1250.

Howe, R. W. (1958). A theoretical evaluation of the potential range and importance of *Trogoderma granarium* Everts in North America (Col. Dermestidae). *Proceedings of the Tenth International Congress of Entomology*, **4**, 23–28.

Huey, R. B. and Berrigan, D. (2001). Temperature, demography, and ectotherm fitness. *American Naturalist*, **158**, 204–210.

Huffaker, C. B. and Messenger, P. S. (1964). The concept and significance of natural control. In *Biological control of insect pests and weeds* (ed. P. De Bach), pp. 74–117. Chapman and Hall, London.

Hughes, J. B. (2000). The scale of resource specialization and the distribution and abundance of lycaenid butterflies. *Oecologia*, **123**, 375–383.

Hughes, J. B., Daily, G. C. and Ehrlich, P. R. (1997). Population diversity: its extent and extinction. *Science*, **278**, 689–692.

Hughes, L., Cawsey, E. M. and Westoby, M. (1996). Geographic and climatic range sizes of Australian eucalypts and a test of Rapoport's rule. *Global Ecology and Biogeography Letters*, **5**, 128–142.

Hugueny, B. (1990). Geographic range of west African freshwater fishes: role of biological characteristics and stochastic processes. *Acta Oecologica*, **11**, 351–375.

Hummel, H., Bogaards, R. H., Bachelet, G., Caron, F., Sola, J. C. and Amiard-Triquet, C. (2000). The respiratory performance and survival of the bivalve *Macoma balthica* (L.) at the southern limit of its distribution area: a translocation experiment. *Journal of Experimental Marine Biology and Ecology*, **251**, 85–102.

Hummel, H., Bogaards, R., Bek, T., Polishchuk, L., Amiard-Triquet, C., Bachelet, G., Desprez, M., Strelkov, P., Sukhotin, A., Naumov, A., Dahle, S., Denisenko, S., Gantsevich, M., Sokolov, K. and de Wolf, L. (1997). Sensitivity to stress in the bivalve *Macoma balthica* from the most northern (Arctic) to the most southern (French) populations: low sensitivity in Arctic populations because of genetic adaptations? *Hydrobiologia*, **355**, 127–138.

Hunter, M. J. Jr and Hutchinson, A. (1994). The virtues and shortcomings of parochialism: conserving species that are locally rare, but globally common. *Conservation Biology*, **8**, 1163–1165.

Huntley, B. (1991). How plants respond to climate change: migration rates, individualism and the consequences for plant communities. *Annals of Botany*, **67** (Suppl. 1), 15–22.

Huntley, B. (1994). Plant species' response to climate change: implications for the conservation of European birds. *Ibis*, **137**, S127–S138.

Huntley, B., Bartlein, P. J. and Prentice, I. C. (1989). Climatic control of the distribution and abundance of beech (*Fagus* L.) in Europe and North America. *Journal of Biogeography*, **16**, 551–560.

Huntley, B., Berry, P. M., Cramer, W. and McDonald, A. P. (1995). Modelling present and potential future ranges of some European higher plants using climate response surfaces. *Journal of Biogeography*, **22**, 967–1001.

Huntley, B. and Birks, H. J. B. (1983). *An atlas of past and present pollen maps for Europe: 0–13 000 years ago.* Cambridge University Press, Cambridge.

Hutchins, L. W. (1947). The bases for temperature zonation in geographical distribution. *Ecological Monographs*, **17**, 325–335.

Hutchinson, A. H. (1918). Limiting factors in relation to specific ranges of tolerance of forest trees. *Botanical Gazette*, **96**, 465–493.

Hutchinson, G. E. (1953). The concept of pattern in ecology. *Proceedings of the Academy of Natural Sciences, Philadelphia*, **105**, 1–12.

Ibrahim, K. M., Nichols, R. A. and Hewitt, G. M. (1996). Spatial patterns of genetic variation generated by different forms of dispersal during range expansion. *Heredity*, **77**, 282–291.

Inkinen, P. (1994). Distribution and abundance in British noctuid moths revisited. *Annales Zoologici Fennici*, **31**, 235–243.

IPCC (Intergovernmental Panel on Climate Change) (2001). *Climate change (2001): The scientific basis*. Contribution of working group I to the third assessment report of the intergovernmental panel on climate change (eds. J. T. Houghton, Y. Ding, D. J. Griggs, M. Noguer, P. J. van der Linden, X. Dai, K. Maskell and C. A. Johnson). Cambridge University Press, Cambridge.

Isaaks, E. H. and Srivastava, R. M. (1989). *An introduction to applied geostatistics*. Oxford University Press, Oxford.

IUCN (1993). *Parks for life—report of the 4th World Conference on Natural Parks and Protected Areas*. IUCN, Gland.

IUCN (1994). *IUCN Red list categories*. IUCN-The World Conservation Union, Gland.

IUCN (1996). *IUCN Red list of threatened animals*. IUCN, Gland.

IUCN/SSC Criteria Review Working Group (1999). IUCN Red list criteria review provisional report: draft of the proposed changes and recommendations. *Species*, **31–32**, 43–57.

Iversen, J. (1944). *Viscum, Hedera* and *Ilex* as climate indicators. *Geologiska Foreningens Forhandlingar Stockholm*, **66**, 463–483.

Ives, A. R. and Klopfer, E. D. (1997). Spatial variation in abundance created by stochastic temporal variation. *Ecology*, **78**, 1907–1913.

Jablonski, D. (1986a). Background and mass extinctions: the alternation of macroevolutionary regimes. *Science*, **231**, 129–133.

Jablonski, D. (1986b). Causes and consequences of mass extinctions: a comparative approach. In *Dynamics of extinction* (ed. D. K. Elliot), pp. 183–229. Wiley, New York.

Jablonski, D. (1986c). Larval ecology and macroevolution in marine invertebrates. *Bulletin of Marine Science*, **39**, 565–587.

Jablonski, D. (1987). Heritability at the species level: analysis of geographic ranges of Cretaceous mollusks. *Science*, **238**, 360–363.

Jablonski, D. (1988). Response [to Russell and Lindberg]. *Science*, **240**, 969.

Jablonski, D., Flessa, K. W. and Valentine, J. W. (1985). Biogeography and paleobiology. *Paleobiology*, **11**, 75–90.

Jablonski, D. and Lutz, R. A. (1983). Larval ecology of marine benthic invertebrates: paleobiological implications. *Biological Reviews*, **58**, 21–89.

Jablonski, D. and Raup, D. M. (1995). Selectivity of end-Cretaceous bivalve extinctions. *Science*, **268**, 389–391.

Jablonski, D. and Valentine, J. W. (1990). From regional to total geographic ranges: testing the relationship in Recent bivalves. *Paleobiology*, **16**, 126–142.

Jackson, J. B. C. (1974). Biogeographic consequences of eurytopy and stenotopy among marine bivalves and their evolutionary consequences. *American Naturalist*, **108**, 541–560.

Jacquez, G. M., Maruca, S. and Fortin, M.-J. (2000). From fields to objects: a review of geographic boundary analysis. *Journal of Geographical Systems*, **2**, 221–241.

Jaeger, R. G. (1970). Potential extinction through competition between two species of terrestrial salamanders. *Evolution*, **24**, 632–642.

Jaeger, R. G. (1971). Competitive exclusion as a factor influencing the distributions of two species of terrestrial salamanders. *Ecology*, **52**, 632–637.

Jain, S. K., Rai, K. N. and Singh, R. S. (1981). Population biology of *Avena* XI. Variation in peripheral isolates of *A. barbata. Genetica*, **56**, 213–215.

Jaksic, F. M. (2001). Ecological effects of El Niño in terrestrial ecosystems of western South America. *Ecography*, **24**, 241–250.

Jalas, J. and Suominen, J. (eds) (1972–1994). *Atlas florae Europaeae*. Vols 1–10. The Committee for Mapping the Flora of Europe and Societas Biologica Fennica Vanamo, Helsinki.

Jalas, J., Suominen, J. and Lampinen, R. (eds) (1996). *Atlas florae Europaeae.* Vol. 11. *Cruciferae (Ricotia to Raphanus).* The Committee for Mapping the Flora of Europe and Societas Biologica Fennica Vanamo, Helsinki.

James, F. C. (1970). Geographic size variation in birds and its relationship to climate. *Ecology,* 51, 365–390.

James, F. C., Johnston, R. F., Wamer, N. O., Niemi, G. J. and Boecklen, W. J. (1984). The Grinnellian niche of the wood thrush. *American Naturalist,* 124, 17–30.

James, T. Y., Porter, D., Hamrick, J. L. and Vilgalys, R. (1999). Evidence for limited intercontinental gene flow in the cosmopolitan mushroom, *Schizophyllum commune. Evolution,* 53, 1665–1677.

Janzen, D. H. (1967). Why mountain passes are higher in the tropics. *American Naturalist,* 101, 233–249.

Järvinen, A. (1986). Clutch size of passerines in harsh environments. *Oikos,* 46, 365–371.

Järvinen, A. (1989). Clutch-size variation in Pied Flycatcher *Ficedula hypoleuca. Ibis,* 131, 572–577.

Järvinen, A. and Väisänen, R. A. (1984). Reproduction of pied flycatchers (*Ficedula hypoleuca*) in good and bad breeding seasons in a northern marginal area. *Auk,* 101, 439–450.

Järvinen, O., Sisula, H., Varvio-Aho, S.-L. and Salminen, P. (1976). Genic variation in isolated marginal populations of the Roman Snail, *Helix pomatia* L. *Hereditas,* 82, 101–110.

Järvinen, O. and Väisänen, R. A. (1979). Climatic changes, habitat changes, and competition: dynamics of geographical overlap in two pairs of congeneric bird species in Finland. *Oikos,* 33, 261–271.

Jarvis, A. M. and Robertson, A. (1999). Predicting population sizes and priority conservation areas for 10 endemic Namibian bird species. *Biological Conservation,* 88, 121–131.

Jeffree, C. E. and Jeffree, E. P. (1996). Redistribution of the potential geographical ranges of Mistletoe and Colorado Beetle in Europe in response to the temperature component of climate change. *Functional Ecology,* 10, 562–577.

Jeffree, E. P. (1959). A climatic pattern between latitudes 40° and 70°N and its probable influence on biological distributions. *Proceedings of the Linnean Society of London,* 171, 89–121.

Jeffree, E. P. and Jeffree, C. E. (1994). Temperature and the biogeographical distributions of species. *Functional Ecology,* 8, 640–650.

Jenkins, A. R. and Hockey, P. A. R. (2001). Prey availability influences habitat tolerance: an explanation for the rarity of peregrine falcons in the tropics. *Ecography,* 24, 359–367.

Jenkins, N. L. and Hoffmann, A. A. (1999). Limits to the southern border of *Drosophila serrata*: cold resistance, heritable variation, and trade-offs. *Evolution,* 53, 1823–1834.

Jenkins, N. L. and Hoffmann, A. A. (2000). Variation in morphological traits and trait asymmetry in field *Drosophila serrata* from marginal populations. *Journal of Evolutionary Biology,* 13, 113–130.

Jenkinson, L. S., Davis, A. J., Wood, S., Shorrocks, B. and Lawton, J. H. (1996). Not that simple: global warming and predictions of insect ranges and abundances – results from a model insect assemblage I replicated laboratory ecosystems. *Aspects of Applied Biology,* 45, 343–348.

Johansson, M. E. (1994). Life history differences between central and marginal populations of the clonal aquatic plant *Ranunculus lingua*: a reciprocal transplant experiment. *Oikos,* 70, 65–72.

Johnson, C. N. (1998a). Species extinction and the relationship between distribution and abundance. *Nature*, **394**, 272–274.

Johnson, C. N. (1998b). Rarity in the tropics: latitudinal gradients in distribution and abundance in Australian mammals. *Journal of Animal Ecology*, **67**, 689–698.

Johnson, K. G., Budd, A. F. and Stemann, T. A. (1995). Extinction selectivity and ecology of Neogene Caribbean reef corals. *Paleobiology*, **21**, 52–73.

Johnson, S. A. and Choromankinorris, J. (1992). Reduction in the eastern limit of the range of the Franklin's ground squirrel (*Spermophilus franklinii*). *American Midland Naturalist*, **128**, 325–331.

Johnston, C. A. (1998). *Geographic information systems in ecology*. Blackwell Science, Oxford.

Johst, K. and Brandl, R. (1997). Body size and extinction risk in a stochastic environment. *Oikos*, **78**, 612–617.

Jones, B., Gliddon, C. and Good, J. E. G. (2001). The conservation of variation in geographically peripheral populations: *Lloydia serotina* (Liliaceae) in Britain. *Biological Conservation*, **101**, 147–156.

Jónsdóttir, I. S., Augner, M., Fagerström, T., Persson, H. and Stenström, A. (2000). Genet age in marginal populations of two clonal *Carex* species in the Siberian Arctic. *Ecography*, **23**, 402–412.

Juliano, S. A. (1983). Body size, dispersal ability, and range size in North American species of *Brachinus* (Coleoptera: Carabidae). *Coleopterists Bulletin*, **37**, 232–238.

Kahmen, S. and Poschlod, P. (2000). Population size, plant performance, and genetic variation in the rare plant *Arnica montana* L. in the Rhön, Germany. *Basic and Applied Ecology*, **1**, 43–51.

Kapos, V. (1989). Effects of isolation on the water status of forest patches in the Brazilian Amazon. *Journal of Tropical Ecology*, **5**, 173–185.

Kapos, V., Wandelli, E., Camargo, J. L. and Ganade, G. (1997). Edge-related changes in environment and plant responses due to forest fragmentation in central Amazonia. In *Tropical forest remnants: ecology, management, and conservation of fragmented communities* (eds W. F. Laurance and R. O. Bierregaard, Jr.), pp. 33–44. University of Chicago Press, Chicago.

Kareiva, P. M., Kingsolver, J. G. and Huey, R. B. (eds) (1993). *Biotic interactions and global change*. Sinauer, Sunderland, Massachusetts.

Karron, J. D. (1987). A comparison of levels of genetic polymorphism and self-compatability in geographically restricted and widespread plant congeners. *Evolutionary Ecology*, **1**, 47–58.

Kat, P. W. (1982). The relationship between heterozygosity for enzyme loci and developmental homeostasis in peripheral populations of aquatic bivalves (Unionidae). *American Naturalist*, **119**, 824–832.

Kattan, G. H., Alvarez-López, H. and Giraldo, M. (1994). Forest fragmentation and bird extinctions: San Antonio eighty years later. *Conservation Biology*, **8**, 138–146.

Kauhala, K. (1995). Changes in distribution of the European badger *Meles meles* in Finland during the rapid colonization of the raccoon dog. *Annales Zoologici Fennici*, **32**, 183–191.

Kavanagh, K. and Kellman, M. (1986). Performance of *Tsuga canadensis* (L.) Carr. at the centre and northern edge of its range: a comparison. *Journal of Biogeography*, **13**, 145–157.

Kavanaugh, D. H. (1979a). Rates of taxonomically significant differentiation in relation to geographical isolation and habitat: examples from a study of the Nearctic *Nebria* fauna. In *Carabid beetles: their evolution, natural history and classification* (eds T. L. Erwin, G. E. Ball and A. L. Halpern), pp. 35–57. Junk, The Hague.

Kavanaugh, D. H. (1979b). Investigations on present climatic refugia in North America through studies on the distributions of carbid beetles: concepts, methodology and prospectus. In *Carabid beetles: their evolution, natural history and classification* (eds T. L. Erwin, G. E. Ball and A. L. Halpern), pp. 371–381. Junk, The Hague.

Kavanaugh, D. H. (1985). On wing atrophy in carabid beetles (Coleoptera: Carabidae), with special reference to Nearctic *Nebria*. In *Taxonomy, phylogeny and zoogeography of beetles and ants* (ed. G. E. Ball.), pp. 408–431. Junk, Dordrecht.

Keleher, C. J. and Rahel, F. J. (1996). Thermal limits to salmonid distributions in the Rocky Mountain Region and potential habitat loss due to global warming: a geographic information system (GIS) approach. *Transactions of the American Fisheries Society*, **125**, 1–13.

Kellner, A. and Green, D. M. (1995). Age structure and age at maturity in Fowler's toads, *Bufo woodhousii fowleri*, at their northern range limit. *Journal of Herpetology*, **29**, 485–489.

Kelly, C. K. (1996). Identifying plant functional types using floristic data bases: ecological correlates of plant range size. *Journal of Vegetation Science*, **7**, 417–424.

Kelly, C. K. and Southwood, T. R. E. (1999). Species richness and resource availability: a phylogenetic analysis of insects associated with trees. *Proceedings of the National Academy of Sciences of the USA*, **96**, 8013–8016.

Kelly, C. K. and Woodward, F. I. (1996). Ecological correlates of plant range size: taxonomies and phylogenies in the study of plant commonness and rarity in Great Britain. *Philosophical Transactions of the Royal Society, London B*, **351**, 1261–1269.

Kemp,W. P. (1992). Rangeland grasshopper (Orthoptera: Acrididae) community structure: a working hypothesis. *Environmental Entomology*, **21**, 461–470.

Kendal, W. S. (1992). Fractal scaling in the geographic distribution of populations. *Ecological Modelling*, **64**, 65–69.

Kendeigh, S. C. (1974). *Ecology with special reference to animals and man*. Prentice Hall, Englewood Cliffs, New Jersey.

Kerney, M. (1999). *Atlas of the land and freshwater molluscs of Britain and Ireland*. Harley, Colchester.

Kerr, J. T. (1999). Weak links: 'Rapoport's rule' and large-scale species richness patterns. *Global Ecology and Biogeography*, **8**, 47–54.

Kieckhefer, R. W. and Elliott, N. C. (1989). Effect of fluctuating temperatures on development of immature Russian wheat aphid (Homoptera: Aphididae) and demographic statistics. *Journal of Economic Entomology*, **82**, 119–122.

Kieckhefer, R. W., Elliott, N. C. and Walgenbach, D. D. (1989). Effects of constant and fluctuating temperatures on developmental rates and demographic statistics of the English grain aphid (Homoptera: Aphididae). *Annals of the Entomological Society of America*, **82**, 701–706.

Kiflawi, M., Enquist, B. J. and Jordan, M. A. (2000). Position within the geographic range, relative local abundance and developmental instability. *Ecography*, **23**, 539–546.

Kinnaird, M. F. and O'Brien, T. G. (1998). Ecological effects of wildfire on lowland rainforest in Sumatra. *Conservation Biology*, **12**, 954–956.

Kinne, O. (1970). Invertebrates. Temperature effects. In *Marine ecology, Vol. 1: Environmental factors, Part 1* (ed. O. Kinne), pp. 407–514. Wiley-Interscience, Chichester.

Kirchhofer, A. (1997). The assessment of fish vulnerability in Switzerland based on distribution data. *Biological Conservation*, **80**, 1–8.

Kirkpatrick, M. and Barton, N. H. (1997). Evolution of a species' range. *American Naturalist*, **150**, 1–23.

Klautau, M., Russo, C. A. M., Lazoski, C., Boury-Esnault, N., Thorpe, J. P. and Solé-Cava, A. M. (1999). Does cosmopolitanism result from overconservative systematics? A case study using the marine sponge *Chondrilla nucula*. *Evolution*, **58**, 1414–1422.

Klein, D. R. (1999). The roles of climate and insularity in establishment and persistence of *Rangifer tarandus* populations in the high Arctic. *Ecological Bulletins*, **47**, 96–104.

Kluyver, H. N. and Tinbergen, L. (1953). Territory and regulation of density in titmice. *Archives Néerlandaises de Zoologie*, **10**, 265–289.

Knowlton, N. (1993). Sibling species in the sea. *Annual Review of Ecology and Systematics*, **24**, 189–216.

Koch, C. F. (1980). Bivalve species duration, areal extent and population size in a Cretaceous sea. *Paleobiology*, **6**, 184–192.

Koch, C. F. (1987). Prediction of sample size effects on measured temporal and geographic distribution patterns of species. *Paleobiology*, **13**, 100–107.

Koch, C. F. and Morgan, J. P. (1988). On the expected distribution of species' ranges. *Paleobiology*, **14**, 126–138.

Koenig, W. D. (1984). Geographic variation in clutch size in the northern flicker (*Colaptes auratus*): support for Ashmole's hypothesis. *Auk*, **101**, 698–706.

Koenig, W. D. (1986). Geographical ecology of clutch size variation in North American woodpeckers. *Condor*, **88**, 499–504.

Koenig, W. D. and Haydock, J. (1999). Oaks, acorns, and the geographical ecology of acorn woodpeckers. *Journal of Biogeography*, **26**, 159–165.

Kohlmann, B., Nix, H. and Shaw, D. D. (1988). Environmental predictions and distributional limits of chromosomal taxa in the Australian grasshopper *Caledia captiva* (F.). *Oecologia*, **75**, 483–493.

Kohonencorish, M., Lokki, J., Saura, A. and Sperlich, D. (1985). The genetic load in a northern marginal population of *Drosophila subobscura*. *Hereditas*, **102**, 255–258.

Kokko, H. and Sutherland, W. J. (2001). Ecological traps in changing environments: ecological and evolutionary consequences of a behaviourally mediated Allee effect. *Evolutionary Ecology Research*, **3**, 537–551.

Koleff, P. and Gaston, K. J. (2001). Latitudinal gradients in diversity: real patterns and random models. *Ecography*, **24**, 341–351.

König, C., Weick, F. and Becking, J.-H. (1999). *Owls: a guide to the owls of the world*. Pica Press, Robertsbridge.

Körner, C. (1998). A re-assessment of high elevation treeline positions and their explanation. *Oecologia*, **115**, 445–459.

Kozár, F. (1995). Geographical segregation of scale-insects (Homoptera: Coccoidea) on fruit trees and the role of host plant ranges. *Acta Zoologica Academiae Scientarium Hungaricae*, **41**, 315–325.

Kozár, F. and Dávid, A. N. (1986). The unexpected northward migration of some species of insects in Central Europe and the climatic changes. *Anzeiger für Schädlingskunde, Pflanzenschutz und Umweltschutz*, **59**, 90–94.

Kozłowski, J. and Gawelczyk, A. T. (2002) Why are species' body size distributions usually skewed to the right? *Functional Ecology*, **16**, 419–432.

Kruess, A. and Tscharntke, T. (1994). Habitat fragmentation, species loss, and biological-control. *Science*, **264**, 1581–1584.

Kuittinen, H., Mattila, A. and Savolainen, O. (1997). Genetic variation at marker loci and in quantitative traits in natural populations of *Arabidopsis thaliana*. *Heredity*, **79**, 144–152.

Kulesza, G. (1990). An analysis of clutch-size in New World passerine birds. *Ibis*, **132**, 407–422.

Kullman, L. (1992). Climatically induced regeneration patterns of marginal populations of *Pinus sylvestris* in northern Sweden. *Oecologia Montana*, **1**, 5–10.

Kunin, W. E. (1998). Extrapolating species abundance across spatial scales. *Science*, **281**, 1513–1515.

Kuno, E. (1986). Evaluation of statistical precision and design of efficient sampling for the population estimates based on frequency of sampling. *Researches in Population Ecology*, **28**, 305–319.

Kuno, E. (1991). Sampling and analysis of insect populations. *Annual Review of Entomology*, **36**, 285–304.

Lack, P. (1986). *The atlas of wintering birds in Britain and Ireland*. Poyser, London.

Lagercrantz, U. and Ryman, N. (1990). Genetic structure of Norway spruce (*Picea abies*): concordance of morphological and allozymic variation. *Evolution*, **44**, 38–53.

Lahti, T., Kemppainen, E., Kurtto, A. and Uotila, P. (1991). Distribution and biological characteristics of threatened vascular plants in Finland. *Biological Conservation*, **55**, 299–314.

Lammi, A., Siikamäki, P. and Mustajärvi, K. (1999). Genetic diversity, population size, and fitness in central and peripheral populations of a rare plant *Lychnis viscaria*. *Conservation Biology*, **13**, 1069–1078.

Lande, R. (1993). Risks of population extinction from demographic and environmental stochasticity and random catastrophes. *American Naturalist*, **142**, 911–927.

Lande, R. (1998). Anthropogenic, ecological and genetic factors in extinction. In *Conservation in a changing world* (eds G. M. Mace, A. Balmford and J. R. Ginsberg), pp. 29–51. Cambridge University Press, Cambridge.

Laurance, W. F. (2000). Do edge effects occur over large spatial scales? *Trends in Ecology and Evolution*, **15**, 134–135.

Laurance, W. F. and Bierregaard, R. O. Jr (eds) (1997). *Tropical forest remnants: ecology, management, and conservation of fragmented communities*. Chicago University Press, Chicago.

Laurance, W. F., Delamônica, P., Laurance, S. G., Vasconcelos, H. L. and Lovejoy, T. E. (2000). Rainforest fragmentation kills big trees. *Nature*, **404**, 836.

Laurance, W. F., Laurance, S. G., Ferreira, L. V., Rankin-de Merona, J. M., Gascon, C. and Lovejoy, T. E. (1997). Biomass collapse in Amazonian forest fragments. *Science*, **278**, 1117–1118.

Laurila, T. and Järvinen, O. (1989). Poor predictability of the threatened status of waterfowl by life-history traits. *Ornis Fennica*, **66**, 165–167.

Lavie, B., Achituv, Y. and Nevo, E. (1993). The niche-width variation hypothesis reconfirmed: validation by genetic diversity in the sessile intertidal cirripedes *Chthamalus stellatus* and *Euraphia depressa* (Crustacea, Chthamalidae). *Zeitschrift fur Zoologische Systematik und Evolutionsforschung*, **31**, 110–118.

Lawton, J. H. (1992). There are not 10 million kinds of population dynamics. *Oikos*, **63**, 337–338.

Lawton, J. H. (1993). Range, population abundance and conservation. *Trends in Ecology and Evolution*, **8**, 409–413.

Lawton, J. H. (1995). The response of insects to environmental change. In *Insects in a changing environment* (eds R. Harrington and N. E. Stork), pp. 3–26. Academic Press, London.

Lawton, J. H. (2000). *Community ecology in a changing world.* Ecology Institute, Oldendorf/Luhe.

Lawton, J. H., MacGarvin, M. and Heads, P. A. (1987). Effects of altitude on the abundance and species richness of insect herbivores on bracken. *Journal of Animal Ecology*, **56**, 147–160.

Lawton, J. H. and Price, P. W. (1979). Species richness of parasites on hosts: agromyzid flies on the British Umbelliferae. *Journal of Animal Ecology*, **48**, 619–637.

Lawton, J. H. and Schröder, D. (1977). Effects of plant type, size of geographical range and taxonomic isolation on numbers of insect species associated with British trees. *Nature*, **265**, 137–140.

Lawton, J. H. and Woodroffe, G. L. (1991). Habitat and the distribution of water voles: why are there gaps in a species' range? *Journal of Animal Ecology*, **60**, 79–91.

Lazarus, D., Hilbrecht, H., Spencer-Cervato, C. and Thierstein, H. (1995). Sympatric speciation and phyletic change in *Globorotalia truncatulinoides*. *Paleobiology*, **21**, 28–51.

Leemans, R. (1996). Biodiversity and global change. In *Biodiversity: a biology of numbers and difference* (ed. K. J. Gaston), pp. 367–387. Blackwell Science, Oxford.

Leggett, W. C. and Frank, K. T. (1997). A comparative analysis of recruitment variability in North Atlantic flatfishes—testing the species range hypothesis. *Journal of Sea Research*, **37**, 281–299.

Leigh, E. G. (1981). The average lifetime of a population in a varying environment. *Journal of Theoretical Biology*, **90**, 231–239.

Leith, H. and Werger, M. J. A. eds. (1989). *Ecosystems of the world 14B: Tropical rain forest ecosystems*. Elsevier, Amsterdam.

Leitner, W. A. and Rosenzweig, M. (1997). Nested species-area curves and stochastic sampling: a new theory. *Oikos*, **79**, 503–512.

Lema, K. M. and Herren, H. R. (1985). The influence of constant temperature on population growth rates of the cassava mealybug, *Phenacoccus manihoti*. *Entomologia Experimentalis et Applicata*, **38**, 165–169.

Lennon, J. J. (2000). Red-shifts and red herrings in geographical ecology. *Ecography*, **23**, 101–113.

Lennon, J. J., Turner, J. R. G. and Connell, D. (1997). A metapopulation model of species boundaries. *Oikos*, **78**, 486–502.

Lesica, P. and Allendorf, F. W. (1995). When are peripheral populations valuable for conservation? *Conservation Biology*, **9**, 753–760.

Letcher, A. J. and Harvey, P. H. (1994). Variation in geographical range size among mammals of the Palearctic. *American Naturalist*, **144**, 30–42.

Letcher, A. J., Purvis, A., Nee, S. and Harvey, P. H. (1994). Patterns of overlap in the geographic ranges of Palearctic and British mammals. *Journal of Animal Ecology*, **63**, 871–879.

Levin, D. A. (1977). The organization of genetic variability in *Phlox drummondii*. *Evolution*, **31**, 477–494.

Levin, D. A. (1979). Genetic variation in annual *Phlox*: self-compatible versus self-incompatible species. *Evolution*, **32**, 245–263.

Levin, D. A. (1993). Local speciation in plants: the rule not the exception. *Systematic Botany*, **18**, 197–208.

Levin, D. A. and Clay, K. (1984). Dynamics of synthetic *Phlox drummondii* populations at the species margin. *American Journal of Botany*, **71**, 1040–1050.

Levin, S. (1992). The problem of pattern and scale in ecology. *Ecology*, **73**, 1943–1967.

Levins, R. (1968). *Evolution in changing environments*. Princeton University Press, Princeton, New Jersey.

Levins, R. (1969). Some demographic and genetic consequences of environmental heterogeneity for biological control. *Bulletin of the Entomological Society of America*, **15**, 237–240.

Levinton, J. (1988). *Genetics, paleontology, and macroevolution*. Cambridge University Press, Cambridge.

Lewington, I., Alström, P. and Colston, P. (1991). *A field guide to the rare birds of Britain and Europe*. HarperCollins, London.

Lewontin, R. C. and Birch, L. C. (1966). Hybridization as a source of variation for adaptation to new environments. *Evolution*, **20**, 315–336.

Li, P. and Adams, W. T. (1989). Range-wide patterns of allozyme variation in Douglas-fir (*Pseudotsuga menziesii*). *Canadian Journal of Forestry Research*, **19**, 149–161.

Lieberman, B. S., Allmon, W. D. and Eldredge, N. (1993). Levels of selection and macroevolutionary patterns in turritellid gastropods. *Paleobiology*, **19**, 205–215.

Liebherr, J. K. and Hajek, A. E. (1990). A cladistic test of the taxon cycle and taxon pulse hypothesis. *Cladistics*, **6**, 39–59.

Linder, E. T., Villard, M.-A., Maurer, B. A. and Scmidt, E. V. (2000). Geographic range structure in North American landbirds: variation with migratory strategy, trophic level, and breeding habitat. *Ecography*, **23**, 678–686.

Lindgren, E., Talleklint, L. and Polfeldt, T. (2000). Impact of climatic change on the northern range limit and population density of the disease-transmitting European tick *Ixodes ricinus*. *Environmental Health Perspectives*, **108**, 119–123.

Lindstedt, S. L. and Boyce, M. S. (1985). Seasonality, fasting endurance, and body size in mammals. *American Naturalist*, **125**, 873–878.

Lindstedt, S. L., Miller, B. J. and Buskirk, S. W. (1986). Home range, time, and body size in mammals. *Ecology*, **67**, 413–418.

Lockwood, J. L., Moulton, M. P. and Balent, K. L. (1999). Introduced avifaunas as natural experiments in community assembly. In *Ecological assembly rules: perspectives, advances, retreats* (eds E. Weiher and P. Keddy), pp. 108–129. Cambridge University Press, Cambridge.

Loehle, C. (1998). Height growth rate tradeoffs determine northern and southern range limits for trees. *Journal of Biogeography*, **25**, 735–742.

Logan, J. A., Wollkind, D. J., Hoyt, S. C. and Tanigoshi, L. K. (1976). An analytical model for description of temperature dependent rate phenomena in arthropods. *Environmental Entomology*, **5**, 1133–1140.

Loik, M. E. and Nobel, P. S. (1993). Freezing tolerance and water relations of *Opuntia fragilis* from Canada and the United States. *Ecology*, **74**, 1722–1732.

Lomolino, M. V. and Channell, R. (1995). Splendid isolation: patterns of geographic range collapse in endangered mammals. *Journal of Mammalogy*, **76**, 335–347.

Lomolino, M. V. and Channell, R. (1998). Range collapse, re-introductions, and biogeographic guidelines for conservation. *Conservation Biology*, **12**, 481–484.

Losos, J. B. (1992). A critical comparison of the taxon-cycle and character-displacement models for size evolution of lizards in the Lesser Antilles. *Copeia*, 1992, 279–288.

Lovejoy, T. E., Bierregaard, R. O. Jr., Rylands, A. B., Malcolm, J. R., Quintela, C. E., Harper, L. H., Brown, K. S. Jr., Powell, A. H., Powell, G. V. N., Schubart, H. O. R. and Hays, M. B. (1986). Edge and other effects of isolation on Amazon forest fragments.

In *Conservation biology: the science of scarcity and diversity* (ed. M. E. Soulé), pp. 257–285. Sinauer, Sunderland, Massachusetts.

Lowe, W. H. and Hauer, F. R. (1999). Ecology of two large, net-spinning caddisfly species in a mountain stream: distribution, abundance, and metabolic response to a thermal gradient. *Canadian Journal of Zoology*, **77**, 1637–1644.

Luijten, S. H., Dierick, A., Gerard, J., Oostermeijer, B., Raijmann, L. E. L. and den Nijs, H. C. M. (2000). Population size, genetic variation, and reproductive success in a rapidly declining, self-incompatible perennial (*Arnica montana*) in The Netherlands. *Conservation Biology*, **14**, 1776–1787.

Luijten, S. H., Oostermeijer, J. G. B., Ellis-Adam, A. C. and den Nijs, H. C. M. (1998). Reproductive biology of the rare biennial *Gentianella germanica* compared with other gentians of different life history. *Acta Botanica Neerlandica*, **47**, 325–336.

Lutz, F. E. (1921). Geographic average, a suggested method for the study of distribution. *American Museum Novitates*, **5**, 1–7.

Lutz, F. E. (1922). Altitude in Colorado and geographical distribution. *Bulletin of American Museum of Natural History*, **96**, 335–366.

Lynch, J. D. (1988). Refugia. In *Analytical biogeography: an integrated approach to the study of animal and plant populations* (ed. A. A. Myers and P. S. Giller), pp. 311–342. Chapman and Hall, London.

Lynch, J. D. (1989). The gauge of speciation: on the frequencies of modes of speciation. In *Speciation and its consequences* (eds D. Otte and J. A. Endler), pp. 527–553. Sinauer, Sunderland, Massachusetts.

Lynch, M. and Lande, R. (1998). The critical effective size for a genetically secure population. *Animal Conservation*, **1**, 70–72.

Lyons, S. K. and Willig, M. R. (1997). Latitudinal patterns of range size: methodological concerns and empirical evaluations for New World bats and marsupials. *Oikos*, **79**, 568–580.

MacArthur, R. H. (1972). *Geographical ecology: patterns in the distribution of species.* Princeton University Press, Princeton, New Jersey.

MacArthur, R. H. and Connell, J. H. (1966). *The biology of populations.* Wiley, New York.

MacCall, A. D. (1990). *Dynamic geography of marine fish populations.* University of Washington Press, Seattle.

Mace, G. M. (1994). An investigation into methods for categorizing the conservation status of species. In *Large-scale ecology and conservation biology* (eds P. J. Edwards, R. M. May and N. R. Webb), pp. 293–312. Blackwell Scientific, Oxford.

Mace, G., Collar, N., Cooke, J., Gaston, K., Ginsberg, J., Leader-Williams, N., Maunder, M. and Milner-Gulland, E. J. (1992). The development of new criteria for listing species on the IUCN red list. *Species*, **19**, 16–22.

Machado-Allison, C. E. and Craig, G. B. Jr (1972). Geographic variation in resistance to desiccation in *Aedes aegypti* and *A. atropalpus* (Diptera: Culicidae). *Annals of the Entomological Society of America*, **65**, 542–547.

Mack, R. N. (1997). Plant invasions: early and continuing expressions of global change. In *Past and future rapid environmental changes: the spatial and evolutionary responses of terrestrial biota* (ed. B. Huntley *et al.*), pp. 205–216. Springer-Verlag, Berlin.

Macpherson, E. (1989). Influence of geographical distribution, body size and diet on population density of benthic fishes off Namibia (South West Africa). *Marine Ecology Progress Series*, **50**, 295–299.

Macpherson, E. and Duarte, C. M. (1994). Patterns in species richness, size and latitudinal range of East Atlantic fishes. *Ecography*, **17**, 242–248.

Malmqvist, B. (2000). How does wing length relate to distribution patterns of stoneflies (Plecoptera) and mayflies (Ephemeroptera)? *Biological Conservation*, **93**, 271–276.

Marchant, J. H., Hudson, R., Carter, S. P. and Whittington, P. (1990). *Population trends in British breeding birds*. British Trust for Ornithology, Thetford.

Margules, C. R. and Pressey, R. L. (2000). Systematic conservation planning. *Nature*, **405**, 243–253.

Markgraf, V., McGlone, M. and Hope, G. (1995). Neogene paleoenvironmental and paleoclimatic change in southern temperate ecosystems—a southern perspective. *Trends in Ecology and Evolution*, **10**, 143–147.

Marquet, P. A. and Taper, M. L. (1998). On size and area: patterns of mammalian body size extremes across landmasses. *Evolutionary Ecology*, **12**, 127–139.

Marren, P. (1999). *Britain's rare flowers*. Poyser, London.

Marshall, C. R. (1991). Estimation of taxonomic ranges from the fossil record. In *Analytical paleobiology* (eds N. Gilinsky and P. Signor), pp. 19–38. Paleontological Society, Knoxville, Tennessee.

Marshall, C. R. (1994). Confidence intervals on stratigraphic ranges: partial relaxation of the assumption of randomly distributed fossil horizons. *Paleobiology*, **20**, 459–469.

Marshall, C. T. and Frank, K. T. (1994). Geographic responses of groundfish to variation in abundance: methods of detection and their interpretation. *Canadian Journal of Fisheries and Aquatic Sciences*, **51**, 808–816.

Marshall, C. T. and Frank, K. T. (1995). Density-dependent habitat selection by juvenile haddock (*Melanogrammus aeglefinus*) on the southwestern Scotian shelf. *Canadian Journal of Fisheries and Aquatic Sciences*, **52**, 1007–1017.

Marshall, J. A. and Haes, E. C. M. (1990). *Grasshoppers and allied insects of Great Britain and Ireland*. Harley, Colchester.

Marshall, J. K. (1968). Factors limiting the survival of *Corynephorus canescens* (L.) Beauv. in Great Britain at the northern edge of its distribution. *Oikos*, **19**, 206–216.

Marti, C. D. (1997). Lifetime reproductive success in barn owls near the limit of the species' range. *Auk*, **114**, 581–592.

Martin, A. P. and Bermingham, E. (2000). Regional endemism and cryptic species revealed by molecular and morphological analysis of a widespread species of Neotropical catfish. *Proceedings of the Royal Society, London B*, **267**, 1135–1141.

Martin, P. S. (2001). Mammals (late Quaternary), extinctions of. In *Encyclopedia of biodiversity, Vol. 3* (ed. S. A. Levin), pp. 825–839. Academic Press, San Diego.

Marzluff, J. M. and Dial, K. P. (1991). Life history correlates of taxonomic diversity. *Ecology*, **72**, 428–439.

Maschiniski, J. (2001). Impacts of ungulate herbivores on a rare willow at the southern edge of its range. *Biological Conservation*, **101**, 119–130.

Matsumura, M. (1992). Life-tables of the migrant skipper, *Parnara guttata guttata* Bremer et Gray (Lepidoptera, Hesperiidae) in the northern peripheral area of its distribution. *Applied Entomology and Zoology*, **27**, 331–340.

Matthews, J. R. (1937). Geographical relationships of the British flora. *Journal of Ecology*, **25**, 1–90.

Mattson, D. J. and Reid, M. M. (1991). Conservation of the Yellowstone grizzly bear. *Conservation Biology*, **5**, 364–372.

Maurer, B. A. (1994). *Geographical population analysis: tools for the analysis of biodiversity*. Blackwell Scientific, Oxford.

Maurer, B. A. (1999). *Untangling ecological complexity*. University of Chicago Press, Chicago.

Maurer, B. A. and Brown, J. H. (1989). Distributional consequences of spatial variation in local demographic processes. *Annales Zoologici Fennici*, **26**, 121–131.

Maurer, B. A., Brown, J. H. and Rusler, R. D. (1992). The micro and macro in body size evolution. *Evolution*, **46**, 939–953.

Maurer, B. A., Ford, H. A. and Rapoport, E. H. (1991). Extinction rate, body size, and avifaunal diversity. *Acta XX Congressus Internationalis Ornithologici*, 826–834.

Maurer, B. A., Linder, E. T. and Gammon, D. (2001). A geographical perspective on the biotic homogenization process: implications from the macroecology of North American birds. In *Biotic homogenization* (eds J. L. Lockwood and M. L. McKinney), pp. 157–177. Kluwer Academic/Plenum, New York.

Maurer, B. A. and Villard, M.-A. (1994). Population density. *National Geographic Research and Exploration*, **10**, 306–317.

May, R. M. (1978). The dynamics and diversity of insect faunas. In *Diversity of insect faunas* (eds L. A. Mound and N. Waloff), pp. 188–204. Blackwell Scientific Publications, Oxford.

May, R. M., Lawton, J. H. and Stork, N. E. (1995). Assessing extinction rates. In *Extinction rates* (eds J. H. Lawton and R. M. May), pp. 1–24. Oxford University Press, Oxford.

Mayfield, H. F. (1983). Kirtland's warbler, victim of its rarity? *Auk*, **100**, 974–976.

Mayr, E. (1963). *Animal species and evolution*. Harvard University Press, Cambridge, Massachusetts.

Mayr, E. (1970). *Populations, species and evolution*. Harvard University Press, Cambridge, Mass.

Mayr, E. (1988). *Toward a new philosophy of biology: observations of an evolutionist*. Harvard University Press, Cambridge, Massachusetts.

McAllister, D. E., Platania, S. P., Schueler, F. W., Baldwin, M. E. and Lee, D. S. (1986). Ichthyofaunal patterns on a geographical grid. In *Zoogeography of freshwater fishes of North America* (ed. C. H. Hocutt and E. D. Wiley), pp. 17–51. Wiley, New York.

McAllister, D. E., Schueler, F. W., Roberts, C. M. and Hawkins, J. P. (1994). Mapping and GIS analysis of the global distribution of coral reef fishes on an equal-area grid. In *Mapping the diversity of nature* (ed. R. I. Miller), pp. 155–175. Chapman and Hall, London.

McArdle, B. H. and Gaston, K. J. (1995). The temporal variability of densities: back to basics. *Oikos*, **74**, 165–171.

McClure, M. S. and Price, P. W. (1976). Ecotope characteristics of coexisting *Erythroneura* leafhoppers (Homoptera; Cicadellidae) on sycamore. *Ecology*, **57**, 928–940.

McCusker, M. R., Parkinson, E. and Taylor, E. B. (2000). Mitochondrial DNA variation in rainbow trout (*Oncorhynchus mykiss*) across its native range: testing biogeographical hypotheses and their relevance to conservation. *Molecular Ecology*, **9**, 2089–2108.

McDonnell, M. J. and Pickett, S. T. A. (1990). Ecosystem structure and function along urban–rural gradients: an unexploited opportunity for ecology. *Ecology*, **71**, 1232–1237.

McGeoch, M. A. and Gaston, K. J. (2000). Edge effects on the prevalence and mortality factors of *Phytomyza ilicis* (Diptera, Agromyzidae) in a suburban woodland. *Ecology Letters*, **3**, 23–29.

McKinney, M. L. (1997). How do rare species avoid extinction? A paleontological view. In *The biology of rarity: causes and consequences of rare-common differences* (eds W. E. Kunin and K. J. Gaston), pp. 110–129. Chapman and Hall, London.

McKinney, M. L. (1998). Is marine biodiversity at less risk? Evidence and implications. *Diversity and Distributions*, **4**, 3–8.

McLaughlin, S. P. (1992). Are floristic areas hierarchically arranged? *Journal of Biogeography*, **19**, 21–32.

McNab, B. K. (1963). Bioenergetics and the determination of home range size. *American Naturalist*, **97**, 133–139.

Meents, J. K., Rice, J., Anderson, B. W. and Ohmart, R. D. (1983). Nonlinear relationships between birds and vegetation. *Ecology*, **64**, 1022–1027.

Mehlman, D. W. (1994). Rarity in North American passerine birds. *Conservation Biology*, **8**, 1141–1145.

Mehlman, D. W. (1997). Change in avian abundance across the geographic range in response to environmental change. *Ecological Applications*, **7**, 614–624.

Meilleur, A., Brisson, J. and Bouchard, A. (1997). Ecological analyses of the northernmost population of pitch pine (*Pinus rigida*). *Canadian Journal of Forest Research*, **27**, 1342–1350.

Meisner, J. D. (1990). Effect of climatic warming on the southern margins of the native range of brook trout, *Salvelinus fontinalis. Canadian Journal of Fisheries and Aquatic Science*, **47**, 1065–1070.

Ménendez, R. and Thomas, C. D. (2000). Metapopulation structure depends on spatial scale in the host-specific moth *Wheeleria spilodactylus* (Lepidoptera: Pterophoridae). *Journal of Animal Ecology*, **69**, 935–951.

Menges, E. S. and Donlan, R. W. (1998). Demographic viability of populations of *Silene regia* in midwestern prairies: relationships with fire management, genetic variation, geographic location, population size and isolation. *Journal of Ecology*, **86**, 63–78.

Merilä, J., Björjlund, M. and Baker, A. J. (1997). Historical demography and present day population structure of the greenfinch, *Carduelis chloris*—an analysis of mtDNA control-region sequences. *Evolution*, **51**, 946–956.

Merriam, C. H. (1894). Laws of temperature control of the geographic distribution of terrestrial animals and plants. *National Geographic*, **6**, 229–238.

Merritt, R., Moore, N. W. and Eversham, B. C. (1996). *Atlas of the dragonflies of Britain and Ireland*. HMSO, London.

Metcalf, A. E., Nunney, L. and Hyman, B. C. (2001). Geographic patterns of genetic differentiation within the restricted range of the endangered Stephens' kangaroo rat *Dipodomys stephensi. Evolution*, **55**, 1233–1244.

Michels, G. J. Jr and Behle, R. W. (1989). Influence of temperature on reproduction, development, and intrinsic rate of increase of Russian wheat aphid, greenbug, and bird cherry-oat aphid (Homoptera: Aphididae). *Journal of Economic Entomology*, **82**, 439–444.

Millar, C. I. and Marshall, K. A. (1991). Allozyme variation of Port Orford Cedar (*Chamaecyparis lawsoniana*)—implications for conservation. *Forest Science*, **37**, 1060–1077.

Miller, A. I. (1997). A new look at age and area: the geographic and environmental expansion of genera during the Ordovician Radiation. *Paleobiology*, **23**, 410–419.

Miller, J. M., Burke, J. S. and Fitzhugh, G. R. (1991). Early life history patterns of Atlantic North American flatfish: likely (and unlikely) factors controlling recruitment. *Netherlands Journal of Sea Research*, **27**, 261–275.

Minchin, P. R. (1989). Montane vegetation of the Mt. Field massif, Tasmania: a test of some hypotheses about properties of community patterns. *Vegetatio*, **83**, 97–110.

Mitchell-Jones, A. J., Amori, G., Bogdanowicz, W., Kryštufek, B., Reijnders, P. J. H., Spitzenberger, F., Stubbe, M., Thissen, J. B. M., Vohralík, V. and Zima, J. (1999). *The atlas of European mammals*. Poyser, London.

Mladenoff, D. J., Sickley, T. A. and Wydeven, A. P. (1999). Predicting gray wolf landscape recolonization: logistic regression models vs. new field data. *Ecological Applications*, **9**, 37–44.

Mockford, S. W., Snyder, M. and Herman, T. B. (1999). A preliminary examination of genetic variation in a peripheral population of Blanding's turtle, *Emydoidea blandingii*. *Molecular Ecology*, **8**, 323–327.

Monroe, B. L. Jr and Sibley, C. G. (1993). *A world checklist of birds*. Yale University Press, New Haven.

Moore, J. A. (1949). Geographic variation of adaptive characters in *Rana pipiens* Schreber. *Evolution*, **3**, 1–24.

Morita, K. and Yamamoto, S. (2000). Occurrence of a deformed white-spotted charr, *Salvelinus leucomaenis* (Pallas), population on the edge of its distribution. *Fisheries Management and Ecology*, **7**, 551–553.

Moulton, M. P. and Pimm, S. L. (1986). Species introduction to Hawaii. In *Ecology of biological invasions of North America and Hawaii* (eds H. A. Mooney and J. A. Drake), pp. 231–249. Springer-Verlag, New York.

Mourelle, C. and Ezcurra, E. (1997). Rapoport's rule: a comparative analysis between South and North American columnar cacti. *American Naturalist*, **150**, 131–142.

Murawski, S. A. (1993). Climate change and marine fish distributions: forecasting from historical analogy. *Transactions of the American Fisheries Society*, **122**, 647–658.

Murcia, C. (1995). Edge effects in fragmented forests: implications for conservation. *Trends in Ecology and Evolution*, **10**, 58–62.

Murphy, S. D. and Vasseur, L. (1995). Pollen limitation in a northern population of *Hepatica acutiloba*. *Canadian Journal of Botany*, **73**, 1234–1241.

Murray, B. R. and Dickman, C. R. (2000). Relationships between body size and geographical range size among Australian mammals: has human impact distorted macroecological patterns? *Ecography*, **23**, 92–100.

Murray, B. R., Fonseca, C. R. and Westoby, M. (1998). The macroecology of Australian frogs. *Journal of Animal Ecology*, **67**, 567–579.

Murray, B. R., Rice, B. L., Keith, D. A., Myerscough, P. J., Howell, J., Floyd, A. G., Mills, K. and Westoby, M. (1999). Species in the tail of rank-abundance curves. *Ecology*, **80**, 1806–1816.

Murray, R. D., Holling, M., Dott, H. E. M. and Vandome, P. (1998). *The breeding birds of south-east Scotland: a tetrad atlas 1988–1994*. Scottish Ornithologists' Club, Edinburgh.

Myers, R. A. (1991). Population variability and range of a species. *Northwest Atlantic Fisheries Organisation Scientific Council Studies*, **16**, 21–24.

Nachman, G. (1981). A mathematical model of the functional relationship between density and spatial distribution of a population. *Journal of Animal Ecology*, **50**, 453–460.

Nachman, G. (1984). Estimates of mean population density and spatial distribution of *Tetranychus urticae* (Acarina: Tetranychidae) and *Phytoseiulus persimilis* (Acarina: Phytoseiidae) based upon the proportion of empty sampling units. *Journal of Applied Ecology*, **21**, 903–913.

Nafus, D. M. (1993). Extinction, biological control, and insect conservation on islands. In *Perspectives on insect conservation* (eds K. J. Gaston, T. R. New and M. J. Samways), pp. 139–154. Intercept, Andover.

Nakano, S., Kitano, F. and Maekawa, K. (1996). Potential fragmentation and loss of thermal habitats for charrs in the Japanese archipelago due to climatic warming. *Freshwater Biology*, **36**, 711–722.

Nantel, P. and Gagnon, D. (1999). Variability in the dynamics of northern peripheral versus southern populations of two clonal plant species, *Helianthus divaricatus* and *Rhus aromatica*. *Journal of Ecology*, **87**, 748–760.

Nash, D. R., Agassiz, D. J. L., Godfray, H. C. J. and Lawton, J. H. (1995). The pattern of spread of invading species: two leaf-mining moths colonizing Great Britain. *Journal of Animal Ecology*, **64**, 225–233.

Nathan, R., Safriel, U. N. and Shirihai, H. (1996). Extinction and vulnerability to extinction at distribution peripheries: an analysis of the Israeli breeding avifauna. *Israel Journal of Zoology*, **42**, 361–383.

National Aeronautics and Space Administration (1988). *Earth system science*. NASA, Washington, D. C.

Neilson, R. P. and Wullstein, L. H. (1983). Biogeography of two southwest American oaks in relation to atmospheric dynamics. *Journal of Biogeography*, **10**, 275–297.

Newmark, W. D. (1987). A land-bridge island perspective on mammalian extinctions in western North American parks. *Nature*, **325**, 430–432.

Newmark, W. D. (1996). Insularization of Tanzanian parks and the local extinction of large mammals. *Conservation Biology*, **10**, 1549–1556.

Newton, A. C. and Haigh, J. M. (1998). Diversity of ectomycorrhizal fungi in Britain: a test of the species-area relationship, and the role of host specificity. *New Phytologist*, **138**, 619–627.

Newton, I. (1995). Relationship between breeding and wintering ranges in Palearctic-African migrants. *Ibis*, **137**, 241–249.

Newton, I. (1997). Links between abundance and distribution of birds. *Ecography*, **20**, 137–145.

Newton, I. (1998). *Population limitation in birds*. Academic Press, London.

Newton, I., Marquiss, M. and Moss, D. (1977). Spacing of sparrowhawk nesting territories. *Journal of Animal Ecology*, **46**, 425–441.

Ney-Nifle, M. and Mangel, M. (1999). Species-area curves based on geographic range and occupancy. *Journal of Theoretical Biology*, **196**, 327–342.

Nicholls, A. O., Viljoen, P. C., Knight, M. H. and van Jaarsveld, A. S. (1996). Evaluating population persistence of censused and unmanaged herbivore populations from the Kruger National Park, South Africa. *Biological Conservation*, **76**, 57–67.

Nilsson, A. N., Elmberg, J. and Sjöberg, K. (1994). Abundance and species richness patterns of predaceous diving beetles (Coleoptera, Dytiscidae) in Swedish lakes. *Journal of Biogeography*, **21**, 197–206.

Nobre, C. A., Sellers, P. J. and Shukla, J. (1991). Amazonian deforestation and regional climate change. *Journal of Climate*, **4**, 957–988.

Noonan, G. R. (1999). GIS analysis of the biogeography of beetles of the subgenus *Anisodactylus* (Insecta: Coleoptera: Carabidae: genus *Anisodactylus*). *Journal of Biogeography*, **26**, 1147–1160.

Norris, R. D. (1991). Biased extinction and evolutionary trends. *Paleobiology*, **17**, 388–399.

Nowell, K. and Jackson, P. (eds) (1996). *Wild cats: status survey and conservation action plan*. IUCN, Gland.

Obeso, J. R. (1992). Geographic distribution and community structure of bumblebees in the northern Iberian peninsula. *Oecologia*, **89**, 244–252.

O'Connor, R. J. (1987). Organization of avian assemblages—the influence of intraspecific habitat dynamics. In *Organization of communities: past and present* (eds J. H. R. Gee and P. S. Giller), pp. 163–183. Blackwell Scientific, Oxford.

O'Connor, R. J. and Shrubb, M. (1986). *Farming and birds*. Cambridge University Press, Cambridge.

Odland, A., Birks, H. J. B. and Line, J. M. (1995). Ecological optima and tolerances of *Thelypteris limbosperma, Athyrium distentifolium*, and *Matteuccia struthiopteris* along environmental gradients in Western Norway. *Vegetatio*, **120**, 115–129.

Oldfield, S., Lusty, C. and MacKinven, A. (1998). *The world list of threatened trees*. WCMC, Cambridge.

Ortega, J. and Arita, H. T. (1998). Neotropical-Nearctic limits in middle America as determined by distributions of bats. *Journal of Mammalogy*, **79**, 772–783.

Owen, J. and Gilbert, F. S. (1989). On the abundance of hoverflies (Syrphidae). *Oikos*, **55**, 183–193.

Pace, M. L. (1993). Forecasting ecological responses to global change: the need for large-scale comparative studies. In *Biotic interactions and global change* (eds P. M. Kareiva, J. G. Kingsolver and R. B. Huey), pp. 356–363. Sinaeur Associates, Sunderland, Massachusetts.

Packer, L. (1990). Solitary and eusocial nests in a population of *Augochlorella striata* (Provancher) (Hymenoptera; Halictidae) at the northern edge of its range. *Behavioral Ecology and Sociobiology*, **27**, 339–344.

Pagel, M. P., May, R. M. and Collie, A. R. (1991). Ecological aspects of the geographic distribution and diversity of mammalian species. *American Naturalist*, **137**, 791–815.

Palumbi, S. R. (1994). Genetic-divergence, reproductive isolation, and marine speciation. *Annual Review of Ecology and Systematics*, **25**, 547–572.

Palumbi, S. R., Grabowsky, G., Duda, T., Geyer, L. and Tachino, N. (1997). Speciation and population genetic structure in tropical Pacific Sea urchins. *Evolution*, **51**, 1506–1517.

Panelius, S. (1978). The detailed geographical distribution of *Tettigonia cantans* in Finland (Orthoptera, Tettigoniidae). *Notulae Entomologicae*, **58**, 151–157.

Panetta, F. D. and Dodd, J. (1987). Bioclimatic prediction of the potential distribution of skeleton weed *Chondrilla juncea* L. in Western Australia. *Journal of the Australian Institute of Agricultural Science*, **53**, 11–16.

Paradis, E., Baillie, S. R., Sutherland, W. J. and Gregory, R. D. (1998). Patterns of natal and breeding dispersal in birds. *Journal of Animal Ecology*, **67**, 518–536.

Parker, V. (1999). *The atlas of the birds of Sul do Save, Southern Mozambique*. Avian Demography Unit and Endangered Wildlife Trust, Cape Town and Johannesburg.

Parmesan, C. (1996). Climate and species' range. *Nature*, **382**, 765–766.

Parmesan, C., Ryrholm, N., Stefanescu, C., Hill, J. K., Thomas, C. D., Descimon, H., Huntley, B., Kaila, L., Kullberg, J., Tammaru, T., Tennent, W. J., Thomas, J. A. and Warren, M. (1999). Poleward shifts in geographical ranges of butterfly species associated with regional warming. *Nature*, **399**, 579–583.

Parry, M. L. and Carter, T. R. (1985). The effect of climatic variations on agricultural risk. *Climatic Change*, **7**, 95–110.

Payette, S. (1993). The range limit of boreal tree species in Québec-Labrador: an ecological and palaeoecological interpretation. *Review of Palaeobotany and Palynology*, **79**, 7–30.

Payette, S. and Delwaide, A. (1994). Growth of black spruce at its northern range limit in Arctic Quebec, Canada. *Arctic and Alpine Research*, **26**, 174–179.

Peach, W. J., Baillie, S. R. and Balmer, D. E. (1998). Long-term changes in the abundance of passerines in Britain and Ireland as measured by constant effort mist-netting. *Bird Study*, **45**, 257–275.

Peakall, D. B. (1970). The eastern bluebird: its breeding season, clutch size, and nesting success. *Living Bird*, **9**, 239–256.

238 · *References*

Pearce, J. and Ferrier, S. (2001). The practical value of modelling relative abundance of species for regional conservation planning: a case study. *Biological Conservation*, **98**, 33–43.

Pearman, D. (1997). Towards a new definition of rare and scarce plants. *Watsonia*, **21**, 231–251.

Pearson, S. M. (1993). The extent and relative influence of landscape-level factors on wintering bird populations. *Landscape Ecology*, **8**, 3–18.

Peat, H. J. and Fitter, A. H. (1994). Comparative analyses of ecological characteristics of British angiosperms. *Biological Reviews*, **69**, 95–115.

Pedersen, A. A. and Loeschcke, V. (2001). Conservation genetics of peripheral populations of the mygalomorph spider *Atypus affinis* (Atypidae) in northern Europe. *Molecular Ecology*, **10**, 1133–1142.

Peery, M. Z. (2000). Factors affecting interspecies variation in home-range sizes of raptors. *Auk*, **117**, 511–517.

Pekkarinen, A. (1989). The hornet (*Vespa crabro* L.) in Finland and its changing northern limit in northwestern Europe. *Entomologisk Tidskfirt*, **110**, 161–164.

Pekkarinen, A. and Teräs, I. (1993). Zoogeography of *Bombus* and *Psithyrus* in northwestern Europe (Hymenoptera, Apidae). *Annales Zoologici Fennici*, **30**, 187–208.

Penry, H. (1994). *Bird atlas of Botswana*. Natal University Press, Scottsville.

Pepper, J. H. (1938). The effect of certain climatic factors on the distribution of the beet webworm (*Loxostege sticticalis* L.) in North America. *Ecology*, **19**, 565–571.

Pérez-Tris, J., Carbonell, R. and Tellería, J. L. (2000). Abundance distribution, morphological variation and juvenile condition of robins, *Erithacus rubecula* (L.), in their Mediterranean range boundary. *Journal of Biogeography*, **27**, 879–888.

Perring, F. H. and Walters, S. (eds) (1962). *Atlas of the British flora*. Nelson, London.

Perry, J. N. (1981). Taylor's power law for dependence of variance on mean in animal populations. *Applied Statistics*, **30**, 254–263.

Perry, J. N. (1987). Host-parasitoid models of intermediate complexity. *American Naturalist*, **130**, 955–957.

Perry, J. N. (1988). Some models for spatial variability of some animal species. *Oikos*, **51**, 124–130.

Perry, J. N. (1994). Chaotic dynamics can generate Taylor's power law. *Proceedings of the Royal Society, London B*, **257**, 221–226.

Perry, J. N. and Taylor, L. R. (1985). Adès: new ecological families of species-specific frequency distributions that describe repeated spatial samples with an intrinsic power-law variance-mean property. *Journal of Animal Ecology*, **54**, 931–953.

Perry, J. N. and Taylor, L. R. (1986). Stability of real interacting populations in space and time: implications, alternatives and the negative binomial k_c. *Journal of Animal Ecology*, **55**, 1053–1068.

Perry, J. N. and Taylor, L. R. (1988). Families of distributions for repeated samples of animal counts. *Biometrics*, **44**, 881–890.

Perry, J. N. and Woiwod, I. P. (1992). Fitting Taylor's power law. *Oikos*, **65**, 538–542.

Peters, R. H. (1983). *The ecological implications of body size.* Cambridge University Press, Cambridge.

Peters, R. L. and Darling, J. D. S. (1985). The greenhouse effect and nature reserves. *BioScience*, **35**, 707–717.

Peters, R. L. and Lovejoy, T. E. (eds) (1992). *Global warming and biological diversity*. Yale University Press, New Haven.

Peterson, M. S. and van der Kooy, S. J. (1997). Distribution, habitat characterization, and aspects of reproduction of a peripheral population of bluespotted sunfish *Enneacanthus gloriosus* (Holbrook). *Journal of Freshwater Ecology*, **12**, 151–161.

Pettersson, R. B. (1997). Lichens, invertebrates and birds in spruce canopies: impacts of forestry. *Acta Universitatis Agriculturae Sueciae: Silvestria* **16**.

Petrova, N. A., Ilinskaya, N. B. and Kaidanov, L. Z. (1996). Adaptiveness of inversion polymorphism in *Chironomus plumosus* (Diptera, Chironomidae): spatial distribution of inversions over species range. *Genetika*, **32**, 1629–1642.

Pfrender, M. E., Bradshaw, W. E. and Kleckner, C. A. (1998). Patterns in the geographical range sizes of ectotherms in North America. *Oecologia*, **115**, 439–444.

Pheloung, P. C. and Scott, J. K. (1996). Climate-based prediction of *Asparagus asparagoides* and *A. declinatus* distribution in Western Australia. *Plant Protection Quarterly*, **11**, 51–53.

Pheloung, P. C., Scott, J. K. and Randall, R. P. (1996). Predicting the distribution of *Emex* in Australia. *Plant Protection Quarterly*, **11**, 138–140.

Pianka, E. R. (1989). Latitudinal gradients in species diversity. *Trends in Ecology and Evolution*, **4**, 223.

Pielou, E. C. (1977a). The latitudinal spans of seaweed species and their patterns of overlap. *Journal of Biogeography*, **4**, 299–311.

Pielou, E. C. (1977b). *Mathematical ecology*. Wiley, New York.

Pielou, E. C. (1978). Latitudinal overlap of seaweed species: evidence for quasi-sympatric speciation. *Journal of Biogeography*, **5**, 227–238.

Pielou, E. C. (1979). *Biogeography*. Wiley, New York.

Pienkowski, M. W. (1984). Breeding biology and population dynamics of Ringed plovers *Charadrius hiaticula* in Britain and Greenland: nest-predation as a possible factor limiting distribution and timing of breeding. *Journal of Zoology, London*, **202**, 83–114.

Pierson, E. A. and Mack, R. N. (1990). The population biology of *Bromus tectorum* in forests: distinguishing the opportunity for dispersal from environmental restriction. *Oecologia*, **84**, 519–525.

Pigott, C. D. (1970). The response of plants to climate and climatic change. In *The flora of a changing Britain* (ed. F. H. Perring), pp. 32–44. Classey, Faringdon.

Pigott, C. D. (1975). Experimental studies on the influence of climate on the geographical distribution of plants. *Weather*, **30**, 82–90.

Pigott, C. D. (1981). Nature of seed sterility and natural regeneration of *Tilia cordata* near its northern limit in Finland. *Annales Botanici Fennici*, **18**, 255–263.

Pigott, C. D. (1989). Factors controlling the distribution of *Tilia cordata* Mill at the northern limits of its geographical range IV. Estimates ages of the trees. *New Phytologist*, **112**, 117–121.

Pigott, C. D. (1992). Are the distributions of species determined by failure to set seed? In *Fruit and seed production* (eds C. Marshall and J. Grace), pp. 203–216. Cambridge University Press, Cambridge.

Pigott, C. D. and Huntley, J. P. (1978). Factors controlling the distribution of *Tilia cordata* at the northern limits of its geographical range I. Distribution in north-west England. *New Phytologist*, **81**, 429–441.

Pigott, C. D. and Huntley, J. P. (1980). Factors controlling the distribution of *Tilia cordata* at the northern limits of its geographical range II. History in north-west England. *New Phytologist*, **84**, 145–164.

Pigott, C. D. and Huntley, J. P. (1981). Factors controlling the distribution of *Tilia cordata* at the northern limits of its geographical range III. Nature and causes of seed sterility. *New Phytologist*, **87**, 817–839.

Pimentel, D. (2001). Agricultural invasions. In *Encyclopedia of biodiversity*, Vol. 1 (ed. S. A. Levin), pp. 71–85. Academic Press, San Diego.

Pimentel, D., Lach, L., Zuniga, R. and Morrison, D. (2000). Environmental and economic costs of nonindigenous species in the United States. *BioScience*, **50**, 53–65.

Pimm, S. L. (1991). *The balance of nature?: Ecological Issues in the conservation of species and communities*. University of Chicago, Chicago.

Pimm, S. L., Russell, G. J., Gittleman, J. L. and Brooks, T. M. (1995). The future of biodiversity. *Science*, **269**, 347–350.

Pinchera, F., Boitani, L. and Corsi, F. (1997). Application to the terrestrial vertebrates of Italy of a system proposed by IUCN for a new classification of national Red List categories. *Biodiversity and Conservation*, **6**, 959–978.

Pitman, N. C. A., Terborgh, J., Silman, M. R. and Nuñez, V. P. (1999). Tree species distributions in an upper Amazonian forest. *Ecology*, **80**, 2651–2661.

Plant, C. (1988). *The butterflies of the London area*. London Natural History Society, London.

Pollard, E. (1979). Population ecology and change in range of the white admiral butterfly *Ladoga camilla* L. in England. *Ecological Entomology*, **4**, 61–74.

Pollard, E., Moss, D. and Yates, T. J. (1995). Population trends of common British butterflies at monitored sites. *Journal of Applied Ecology*, **32**, 9–16.

Pomeroy, D. and Ssekabiira, D. (1990). An analysis of the distributions of terrestrial birds in Africa. *African Journal of Ecology*, **28**, 1–13.

Poulin, R. (1998). *Evolutionary ecology of parasites: from individuals to communities*. Chapman and Hall, London.

Poulsen, B. O. and Krabbe, N. (1997). Avian rarity in ten cloud-forest communities in the Andes of Ecuador: implications for conservation. *Biodiversity and Conservation*, **6**, 1365–1375.

Powers, D. M. (1979). Evolution in peripheral isolated populations: *Carpodacus* finches on the California islands. *Evolution*, **33**, 834–847.

Pregill, G. K. and Olson, S. L. (1981). Zoogeography of West Indian vertebrates in relation to Pleistocene climatic cycles. *Annual Review of Ecology and Systematics*, **12**, 75–98.

Premoli, A. C. (1997). Genetic variation in a geographically restricted and two widespread species of South American *Nothofagus*. *Journal of Biogeography*, **24**, 883–892.

Pressey, R. L., Humphries, C. J., Margules, C. R., Vane-Wright, R. I. and Williams, P. H. (1993). Beyond opportunism: key principles for systematic reserve selection. *Trends in Ecology and Evolution*, **8**, 124–128.

Price, J., Droege, S. and Price, A. (1995). *The summer atlas of North American birds*. Academic Press, London.

Price, P. W., Westoby, M. and Rice, B. (1988). Parasite-mediated competition: some predictions and tests. *American Naturalist*, **131**, 544–555.

Price, T. D., Helbig, A. J. and Richman, A. D. (1997). Evolution of breeding distributions in the Old World leaf warblers (Genus *Phylloscopus*). *Evolution*, **51**, 552–561.

Primack, R. B. and Miao, S. L. (1992). Dispersal can limit local plant distribution. *Conservation Biology*, **6**, 513–519.

Prince, S. D. (1976). The effect of climate on grain development in barley at an upland site. *New Phytologist*, **76**, 377–389.

Prince, S. D. and Carter, R. N. (1985). The geographical distribution of prickly lettuce (*Lactuca serriola*): III. Its performance in transplant sites beyond its distribution limit in Britain. *Journal of Ecology*, **73**, 49–64.

Prince, S. D., Carter, R. N. and Dancy, K. J. (1985). The geographical distribution of prickly lettuce (*Lactuca serriola* L.): II. Characteristics of populations near its distribution limits in Britain. *Journal of Ecology*, **73**, 39–48.

Pulliam, H. R. (1988). Sources, sinks, and population regulation. *American Naturalist*, **132**, 652–661.

Purvis, A., Gittleman, J. L., Cowlishaw, G. and Mace, G. M. (2000). Predicting extinction risk in declining species. *Proceedings of the Royal Society, London B*, **267**, 1947–1952.

Pyron, M. (1999). Relationships between geographical range size, body size, local abundance, and habitat breadth in North American suckers and sunfishes. *Journal of Biogeography*, **26**, 549–558.

Quinn, R. M. (1995). *The biogeography of rare species and their conservation*. Unpublished PhD thesis. University of London.

Quinn, R. M., Gaston, K. J. and Arnold, H. R. (1996). Relative measures of geographic range size: empirical comparisons. *Oecologia*, **107**, 179–188.

Quinn, R. M., Gaston, K. J., Blackburn, T. M. and Eversham, B. C. (1997b). Abundance-range size relationships of macrolepidoptera in Britain: the effects of taxonomy and life history variables. *Ecological Entomology*, **22**, 453–461.

Quinn, R. M., Gaston, K. J. and Roy, D. B. (1997a). Coincidence between consumer and host occurrence: macrolepidoptera in Britain. *Ecological Entomology*, **22**, 197–208.

Quinn, R. M., Gaston, K. J. and Roy, R. B. (1998). Coincidence in the distributions of butterflies and their foodplants. *Ecography*, **21**, 279–288.

Radomski, A. A., Osborn, D. A., Pence, D. B., Nelson, M. I. and Warren, R. J. (1991). Visceral helminths from an expanding insular population of the long-nosed armadillo (*Dasypus novemcinctus*). *Journal of the Helminthological Society of Washington*, **58**, 1–6.

Ramankutty, N. and Foley, J. A. (1999). Estimating historical changes in land cover: North American croplands from 1850 to 1992. *Global Ecology and Biogeography*, **8**, 381–396.

Ramsar Convention Bureau. (2000). *Ramsar handbooks for the wise use of wetlands*. Ramsar Convention Bureau, Gland, Switzerland.

Randall, M. G. M. (1982). The dynamics of an insect population throughout its altitudinal distribution: *Coleophora alticolella* (Lepidoptera) in northern England. *Journal of Animal Ecology*, **51**, 993–1016.

Randolph, S. E. (1997). Abiotic and biotic determinants of the seasonal dynamics of the tick *Rhipicephalus appendiculatus* in South Africa. *Medical and Veterinary Entomology*, **11**, 25–37.

Randolph, S. E. and Rogers, D. J. (2000). Fragile transmission cycles of tick-borne encephalitis virus may be disrupted by predicted climate change. *Proceedings of the Royal Society, London B*, **267**, 1741–1744.

Rapoport, E. H. (1982). *Areography: geographical strategies of species*. Pergamon, Oxford.

Rapoport, E. H. (1994). Remarks on marine and continental biogeography: an areographical viewpoint. *Philosophical Transactions of the Royal Society, London B*, **343**, 71–78.

Rapoport, E. H. (2000). Remarks on the biogeography of land invasions. *Revista Chilena de Historia Natural*, **73**, 367–380.

Rapoport, E. H., Borioli, G., Monjeau, J. A., *et al.* (1986). The design of nature reserves: a simulation trial for assessing specific conservation value. *Biological Conservation*, **37**, 269–290.

Reaka, M. L. (1980). Geographic range, life history patterns, and body size in a guild of coral-dwelling mantis shrimps. *Evolution*, **34**, 1019–1030.

Reaka-Kudla, M. L., Wilson, D. E. and Wilson, E. O. (eds) (1997). *Biodiversity II: understanding and protecting our biological resources*. Joseph Henry Press, Washington, D. C.

Rees, M. (1995). Community structure in sand dune annuals: is seed weight a key quantity? *Journal of Ecology*, **83**, 857–863.

Reinhardt, K. (1997). Breeding success of southern hemisphere skuas *Catharacta* spp.: the influence of latitude. *Ardea*, **85**, 73–82.

Repasky, R. R. (1991). Temperature and the northern distributions of wintering birds. *Ecology*, **72**, 2274–2285.

Rex, M. A., Etter, R. J. and Stuart, C. T. (1997). Large-scale patterns of species diversity in the deep-sea benthos. In *Marine biodiversity: patterns and processes* (eds R. F. G. Ormond, J. D. Gage and M. V. Angel), pp. 94–121. Cambridge University Press, Cambridge.

Ribera, I. and Vogler, A. P. (2000). Habitat type as a determinant of species range sizes: the example of lotic-lentic differences in aquatic Coleoptera. *Biological Journal of the Linnean Society*, **71**, 33–52.

Ricciardi, A. and Rasmussen, J. B. (1999). Extinction rates of North American freshwater fauna. *Conservation Biology*, **13**, 1220–1222.

Rice, W. R. and Hostert, E. E. (1993). Laboratory experiments on speciation: what have we learned in 40 years? *Evolution*, **47**, 1637–1653.

Richards, O. W. and Southwood, T. R. E. (1968). The abundance of insects: Introduction. In *Insect abundance* (ed. T. R. E. Southwood), pp. 1–7. Blackwell Scientific, Oxford.

Richards, P. W. (1973). Africa, the 'Odd Man Out'. In *Tropical forest ecosystems in Africa and South America: a comparative review* (eds B. J. Meggers, E. S. Ayensu and W. D. Duckworth), pp. 21–26. Smithsonian Institution Press, Washington, D. C.

Richter, T. A., Webb, P. I. and Skinner, J. D. (1997). Limits to the distribution of the southern African Ice Rat (*Otomys sloggetti*): thermal physiology or competitive exclusion? *Functional Ecology*, **11**, 240–246.

Rick, C. M., Forbes, J. F. and Holle, M. (1977). Genetic variation in *Lycopersicon pimpinnellifolium*: evidence of evolutionary change in mating systems. *Plant Systematics and Evolution*, **127**, 139–170.

Ricklefs, R. E. (1972). Latitudinal variation in breeding productivity of the rough-winged swallow. *Auk*, **89**, 826–836.

Ricklefs, R. E. and Bermingham, E. (1999). Taxon cycles in the Lesser Antillean avifauna. *Ostrich*, **70**, 49–59.

Ricklefs, R. E. and Cox, G. W. (1972). Taxon cycles in the West Indies avifauna. *American Naturalist*, **106**, 195–219.

Ricklefs, R. E. and Cox, G. W. (1978). Stage of taxon cycle, habitat distribution and population density in the avifauna of the West Indies. *American Naturalist*, **122**, 875–895.

Ricklefs, R. E. and Latham, R. E. (1992). Intercontinental correlation of geographical ranges suggests stasis in ecological traits of relict genera of temperate perennial herbs. *American Naturalist*, **139**, 1305–1321.

Rijnsdorp, A. D., van Beek, F. A., Flatman, S., Millner, R. M., Riley, J. D., Giret, M. and de Clerck, R. (1992). Recruitment of sole stocks, *Solea solea* (L.), in the northeast Atlantic. *Netherlands Journal of Sea Research*, **29**, 173–192.

Ripley, S. D. and Beehler, B. M. (1990). Patterns of speciation in Indian birds. *Journal of Biogeography*, **17**, 639–648.

Robbins, C. S., Droege, S. and Sauer, J. R. (1989b). Monitoring bird populations with Breeding Bird Survey and atlas data. *Annales Zoologici Fennici*, **26**, 297–304.

Roberts, C. M. and Hawkins, J. P. (1999). Extinction in the sea. *Trends in Ecology and Evolution*, **14**, 241–246.

Robinson, T., Rogers, D. and Williams, B. (1997a). Univariate analysis of tsetse habitat in the common fly belt of Southern Africa using climate and remotely sensed vegetation data. *Medical and Veterinary Entomology*, **11**, 223–234.

Robinson, T., Rogers, D. and Williams, B. (1997b). Mapping tsetse habitat suitability in the common fly belt of Southern Africa using multivariate analysis of climate and remotely sensed vegetation data. *Medical and Veterinary Entomology*, **11**, 235–245.

Rodrigues, A. S. L., Orestes Cerdeira, J. and Gaston, K. J. (2000a). Flexibility, efficiency, and accountability: adapting reserve selection algorithms to more complex conservation problems. *Ecography*, **23**, 565–574.

Rodrigues, A. S. L., Gregory, R. D. and Gaston, K. J. (2000b). Robustness of reserve selection procedures under temporal species turnover. *Proceedings of the Royal Society, London B*, **267**, 49–56.

Rodrigues, A. S. L. and Gaston, K. J. (2001). How large do reserve networks need to be? *Ecological Letters*, **4**, 602–609.

Rodrigues, A. S. L. and Gaston, K. J. (2002). Rarity and conservation planning across geopolitical units. *Conservation Biology*, **16**, 674–682.

Rogers, D. J. (1979). Tsetse population dynamics and distribution: a new analytical approach. *Journal of Animal Ecology*, **48**, 825–849.

Rogers, D. J. and Randolph, S. E. (1986). Distribution and abundance of tsetse flies (*Glossina* spp.). *Journal of Animal Ecology*, **55**, 1007–1025.

Rogers, D. J. and Randolph, S. E. (1991). Mortality rates and population density of tsetse flies correlated with satellite data. *Nature*, **351**, 739–741.

Rogers, D. J. and Randolph, S. E. (2000). The global spread of malaria in a future, warmer world. *Science*, **289**, 1763–1766.

Rogers, D. J. and Williams, B. G. (1994). Tsetse distribution in Africa: seeing the wood and the trees. In *Large-scale ecology and conservation biology* (eds P. J. Edwards, R. M. May and N. R. Webb), pp. 247–271. Blackwell Scientific, Oxford.

Rohde, K. (1996). Rapoport's rule is a local phenomenon and cannot explain latitudinal gradients in species diversity. *Biodiversity Letters*, **3**, 10–13.

Rohde, K. and Heap, M. (1996). Latitudinal ranges of teleost fish in the Atlantic and Indo-Pacific oceans. *American Naturalist*, **147**, 659–665.

Rohde, K., Heap, M. and Heap, D. (1993). Rapoport's rule does not apply to marine teleosts and cannot explain latitudinal gradients in species richness. *American Naturalist*, **142**, 1–16.

Rolstad, J., Loken, B. and Rolstad, E. (2000). Habitat selection as a hierarchical spatial process: the green woodpecker at the northern edge of its distribution range. *Oecologia*, **124**, 116–129.

Romero, J. and Real, R. (1996). Macroenvironmental factors as ultimate determinants of distribution of common toad and natterjack toad in the south of Spain. *Ecography*, **19**, 305–312.

Root, T. (1988a). *Atlas of wintering North American birds*. University of Chicago Press, Chicago.

Root, T. (1988b). Environmental factors associated with avian distributional boundaries. *Journal of Biogeography*, **15**, 489–505.

Root, T. (1988c). Energy constraints on avian distributions and abundances. *Ecology*, **69**, 330–339.

Root, T. (1989). Energy constraints on avian distributions: a reply to Castro. *Ecology*, **70**, 1183–1185.

Root, T. (1991). Positive correlation between range size and body size: a possible mechanisms. *Acta XX Congressus Internationalis Ornithologici*, 817–825.

Root, T. L. and Schneider, S. H. (1993). Can large-scale climatic models be linked with multiscale ecological studies? *Conservation Biology*, **7**, 256–270.

Rose, G. A. and Leggett, W. C. (1991). Effects of biomass-range interactions on catchability of migratory demersal fish by mobile fisheries: an example of Atlantic cod (*Gadus morhua*). *Canadian Journal of Fisheries and Aquatic Sciences*, **48**, 843–848.

Rosen, B. R. (1988). Biogeographic patterns: a perceptual overview. In *Analytical biogeography: an integrated approach to the study of animal and plant distributions* (eds A. A. Myers and P. S. Giller), pp. 23–55. Chapman and Hall, London.

Rosenzweig, M. L. (1975). On continental steady states of species diversity. In *Ecology and evolution of communities* (eds. M. L. Cody and J. M. Diamond), pp. 124–140. Harvard University Press, Cambridge, Massachusetts.

Rosenzweig, M. L. (1978). Geographical speciation: on range size and the probability of isolate formation. In *Proceedings of the Washington State University Conference on Biomathematics and Biostatistics* (ed. D. Wollkind), pp. 172–194. Washington State University, Washington.

Rosenzweig, M. L. (1991). Habitat selection and population interactions: the search for mechanism. *American Naturalist*, **137**, S45–S28.

Rosenzweig, M. L. (1992). Species diversity gradients: we know more and less than we thought. *Journal of Mammalogy*, **73**, 715–730.

Rosenzweig, M. L. (1995). *Species diversity in space and time*. Cambridge University Press, Cambridge.

Rosewell, J., Shorrocks, B. and Edwards, K. (1990). Competition on a divided and ephemeral resource: testing the assumptions. I. Aggregation. *Journal of Animal Ecology*, **59**, 977–1001.

Rossetto, M., Lucarotti, F., Hopper, S. D. and Dixon, K. W. (1997). DNA fingerprinting of *Eucalyptus graniticola*: a critically endangered relict species or a rare hybrid? *Heredity*, **79**, 310–318.

Roughgarden, J. and Pacala, S. (1989). Taxon cycle among *Anolis* lizard populations: review of evidence. In *Speciation and its consequences* (eds. D. Otte and J. A. Endler), pp. 403–432. Sinauer Associates, Sunderland, Massachusetts.

Routledge, R. D. and Swartz, T. B. (1991). Taylor's power law re-examined. *Oikos*, **60**, 107–112.

Rowe, G., Beebee, T. J. C. and Burke, T. (1998). Phylogeography of the natterjack toad *Bufo calamita* in Britain: genetic differentiation of native and translocated populations. *Molecular Ecology*, **7**, 751–760.

Roy, J., Navas, M. L. and Sonié, L. (1991). Invasion by annual brome grasses: a case study challenging the homocline approach to invasions. In *Biogeography of Mediterranean invasions* (eds R. H. Groves and F. Di Castri), pp. 207–224. Cambridge University Press, Cambridge.

Roy, K. (1994). Effects of the Mesozoic Marine Revolution on the taxonomic, morphologic, and biogeographic evolution of a group: aporrhaid gastropods during the Mesozoic. *Paleobiology*, **20**, 274–296.

Roy, K., Jablonski, D. and Valentine, J. W. (1994). Eastern Pacific molluscan provinces and latitudinal diversity gradient: no evidence for Rapoport's rule. *Proceedings of the National Academy of Sciences of the USA*, **91**, 8871–8874.

Roy, K., Jablonski, D. and Valentine, J. W. (1995). Thermally anomalous assemblages revisited: patterns in the extraprovincial latitudinal range shifts of Pleistocene marine mollusks. *Geology*, **23**, 1071–1074.

Roy, K., Jablonski, D., Valentine, J. W. and Rosenberg, G. (1998). Marine latitudinal diversity gradients: tests of causal hypotheses. *Proceedings of the National Academy of Sciences of the USA*, **95**, 3699–3702.

Royama, T. (1969). A model for the global variation of clutch size in birds. *Oikos*, **20**, 562–567.

Ruggiero, A. (1994). Latitudinal correlates of the sizes of mammalian geographical ranges in South America. *Journal of Biogeography*, **21**, 545–559.

Ruggiero, A. and Lawton, J. H. (1998). Are there latitudinal and altitudinal Rapoport effects in the geographic ranges of Andean passerine birds? *Biological Journal of the Linnean Society*, **63**, 283–304.

Ruiz, G. M., Carlton, J. T., Grosholz, E. D. and Hines, A. H. (1997). Global invasions of marine and estuarine habitats by non-indigenous species: mechanisms, extent and consequences. *American Zoologist*, **37**, 621–632.

Rummel, J. D. and Roughgarden, J. (1985). A theory of faunal buildup for competition communities. *Evolution*, **39**, 1009–1033.

Rumsey, F. J., Vogel, J. C., Russell, S. J., Barrett, J. A. and Gibby, M. (1999). Population structure and conservation biology of the endangered fern *Trichomanes speciosum* Willd. (Hymenophyllaceae) at its northern distributional limit. *Biological Journal of the Linnean Society*, **66**, 333–344.

Ruohomäki, K., Kaitaniemi, P., Kozlov, M., Tammaru, T. and Haukioja, E. (1996). Density and performance of *Epirrita autumnata* (Lepidoptera: Geometridae) along three air pollution gradients in northern Europe. *Journal of Applied Ecology*, **33**, 773–785.

Ruokolainen, K. and Vormisto, J. (2000). The most widespread Amazonian palms tend to be tall and habitat generalists. *Basic and Applied Ecology*, **1**, 97–108.

Rushton, S. P., Lurz, P. W. W., Gurnell, J. and Fuller, R. (2000). Modelling the spatial dynamics of parapoxvirus disease in red and grey squirrels: a possible cause of the decline in the red squirrel in the UK? *Journal of Applied Ecology*, **37**, 997–1012.

Russell, M. P. and Lindberg, D. R. (1988a). Real and random patterns associated with molluscan spatial and temporal distributions. *Paleobiology*, **14**, 322–330.

Russell, M. P. and Lindberg, D. R. (1988b). Estimates of species duration. *Science*, **240**, 969.

Rutherford, M. C., O'Callaghan, M., Hurford, J. L., Powrie, L. W., Schulze, R. E., Kunz, R. P., Davis, G. W., Hoffman, M. T. and Mack, F. (1995). Realized niche spaces and functional types: a framework for prediction of compositional change. *Journal of Biogeography*, **22**, 523–531.

Ryrholm, N. (1988). An extralimital population in a warm climatic outpost: the case of the moth *Idaea dilutaria* in Scandinavia. *International Journal of Biometeorology*, **32**, 205–216.

Ryrholm, N. (1989). The influence of the climatic energy balance on living conditions and distribution patterns of *Idaea* spp. (Lepidoptera: Geometridae): an expansion of the species energy theory. *Acta Universitatis Upsaliensis*, **208**, 1–27.

Safriel, U. N., Volis, S. and Kark, S. (1994). Core and peripheral populations and global climate change. *Israel Journal of Plant Sciences*, **42**, 331–345.

Sage, R. D. and Wolff, J. O. (1986). Pleistocene glaciations, fluctuating ranges, and low genetic variability in a large mammal (*Ovis dalli*). *Evolution*, **40**, 1092–1095.

Salisbury, E. J. (1926). The geographical distribution of plants in relation to climatic factors. *Geographical Journal*, **57**, 312–335.

Salisbury, E. J. (1932). The East-Anglian flora. *Transactions of the Norfolk and Norwich Naturalists Society*, **13**, 191–263.

Samways, M. J. (1994). *Insect conservation biology*. Chapman and Hall, London.

Samways, M. J., Osborn, R., Hastings, H. and Hattingh, V. (1999). Global climate change and accuracy of prediction of species' geographical ranges: establishment success of introduced ladybirds (Coccinellidae, *Chilocorus* spp.) worldwide. *Journal of Biogeography*, **26**, 795–812.

Sands, D. P. A., Schotz, M. and Bourne, A. S. (1986). A comparative study on the intrinsic rates of increase of *Cyrtobagous singularis* and *C. salviniae* on the water weed *Salvinia molesta*. *Entomologia Experimentalis et Applicata*, **42**, 231–237.

Santelices, B. and Marquet, P. A. (1998). Seaweeds, latitudinal diversity patterns, and Rapoport's Rule. *Diversity and Distributions*, **4**, 71–75.

Santelmann, M. V. (1991). Influences on the distribution of *Carex exilis*: an experimental approach. *Ecology*, **72**, 2025–2037.

Sanz, J. J. (1997). Geographic variation in breeding parameters of the Pied Flycatcher *Ficedula hypoleuca*. *Ibis*, **139**, 107–114.

Sauer, J. R., Hines, J. E., Gough, G., Thomas, I. and Peterjohn, B. G. (1997). *The North American Breeding Bird Survey results and analysis. Version 96.4*. Patuxent Wildlife Research Center, Laurel, Maryland. http://www.mbr-pwrc.usgs.gov/bbs/bbs.html.

Saunders, D. and Ingram, J. (1995). *Birds of southwestern Australia*. Surrey Beatty, Chipping Norton.

Saunders, D. A., Hobbs, R. J. and Margules, C. R. (1991) Biological consequences of ecosystem fragmentation: a review. *Conservation Biology*, **5**, 18–32.

Scharf, F. S., Juanes, F. and Sutherland, M. (1998). Inferring ecological relationships from the edges of scatter diagrams: comparison of regression techniques. *Ecology*, **79**, 448–460.

Schliewen, U. K., Tautz, D. and Pääbo, S. (1994). Sympatric speciation suggested by monophyly of crater lake cichlids. *Nature*, **368**, 629–632.

Schmitt, T. and Seitz, A. (2001). Allozyme variation in *Polyommatus corridon* (Lepidoptera: Lycaenidae): identification of ice-age refugia and reconstruction of post-glacial expansion. *Journal of Biogeography*, **28**, 1129–1136.

Schnabel, A. and Hamrick, J. L. (1990). Organization of genetic diversity within and among popultions of *Gleditisia triacanthos* (Leguminosae). *American Journal of Botany*, **77**, 1060–1069.

Schneider, C. J., Smith, T. B., Larison, B. and Moritz, C. (1999). A test of alternative models of diversification in tropical rainforests: Ecological gradients vs. rainforest refugia. *Proceedings of the National Academy of Sciences of the USA*, **96**, 13869–13873.

Schoener, T. W. (1968). Sizes of feeding territories among birds. *Ecology*, **49**, 123–141.

Schoener, T. W. (1987). The geographical distribution of rarity. *Oecologia*, **74**, 161–173.

Schonewald-Cox, C. and Buechner, M. (1991). Housing viable populations in protected habitats: the value of a coarse-grained geographic analysis of density patterns and available habitat. In *Species conservation: a population-biological approach* (eds A. Seitz and V. Loeschcke), pp. 213–226. Birkhauser Verlag, Basel.

Schorger, A. W. (1955). *The passenger pigeon: its natural history and extinction.* University of Wisconsin Press, Madison.

Schwaegerle, K. E. and Schaal, B. A. (1979). Genetic variability and founder effect in the pitcher plant *Saracevia purpurea* L. *Evolution*, **33**, 1210–1218.

Sealy, J. R. and Webb, D. A. (1950). Biological flora of the British Isles: *Arbutus nuedo* L. *Journal of Ecology*, **38**, 223–236.

Sepkoski, J. J. and Rex, M. A. (1974). Distribution of freshwater mussels: coastal river as biogeographic islands. *Systematic Zoology*, **23**, 165–188.

Sexton, O. J., Andrews, R. M. and Bramble, J. E. (1992). Size and growth-rate characteristics of a peripheral-population of *Crotaphytus collaris* (Sauria, Crotaphytidae). *Copeia*, 1992, 968–980.

Shao, G. and Halpin, P. N. (1995). Climatic controls of eastern North American coastal tree and shrub distributions. *Journal of Biogeography*, **22**, 1083–1089.

Sharrock J. T. R. (1976). *The atlas of breeding birds of Britain and Ireland.* Poyser, Berkhamsted.

Shaw, A. J. (2001). Biogeographic patterns and cryptic speciation in bryophytes. *Journal of Biogeography*, **28**, 253–261.

Shelford, V. E. (1911). Physiological animal geography. *Journal of Morphology*, **22**, 551–618.

Shorrocks, B. and Rosewell, J. (1986). Guild size in drosophilids: a simulation model. *Journal of Animal Ecology*, **55**, 527–541.

Shreeve, T. G., Dennis, R. L. H. and Pullin, A. S. (1996). Marginality: scale determined processes and the conservation of the British butterfly fauna. *Biodiversity and Conservation*, **5**, 1131–1141.

Shugart, H. H. (1998). *Terrestrial ecosystems in changing environments.* Cambridge University Press, Cambridge.

Shugart, H. H. and Patten, B. C. (1972). Niche quantification and the concept of niche pattern. In *Systems analysis and simulation ecology* (ed. B. C. Patten), pp. 283–327. Academic Press, New York.

Shumaker, K. M. and Babbel, G. R. (1980). Patterns of allozymic similarity in ecologically central and marginal populations of *Hordeum jubatum* in Utah. *Evolution*, **34**, 110–116.

Shuter, B. J. and Post, J. R. (1990). Climate, population viability, and the zoogeography of temperate fishes. *Transactions of the American Fisheries Society*, **119**, 314–336.

Sibley, C. G. and Monroe, B. L. Jr (1990). *Distribution and taxonomy of birds of the World.* Yale University Press, New Haven.

Siikamäki, P. and Lammi, A. (1998). Fluctuating asymmetry in central and marginal populations of *Lychnis viscaria* in relation to genetic and environmental factors. *Evolution*, **52**, 1285–1292.

Silander, J. A. and Antonovics, J. (1982). A perturbation approach to the analysis of interspecific interactions in a coastal plant community. *Nature*, **298**, 557–560.

Silva, M. and Downing, J. A. (1995). *CRC Handbook of mammalian body masses.* CRC Press, Boca Raton.

Simková, A., Morand, S., Matejusová, I., Jurajda, P. and Gelnar, M. (2001). Local and regional influences on patterns of parasite species richness of central European fishes. *Biodiversity and Conservation*, **10**, 511–525.

Siriwardena, G. M., Baillie, S. R., Buckland, S. T., Fewster, R. M., Marchant, J. H. and Wilson, J. D. (1998). Trends in the abundance of farmland birds: a quantitative comparison of smoothed Common Birds Census indices. *Journal of Applied Ecology*, **35**, 24–43.

Sisk, T. D. and Margules, C. R. (1993). Habitat edges and restoration: methods for quantifying edge effects and predicting the results of restoration efforts. In *Nature conservation 3: reconstruction of fragmented ecosystems* (eds D. A. Saunders, R. J. Hobbs and P. R. Ehrlich), pp. 57–68. Surrey Beatty, Sydney.

Sjöberg, M., Albrectsen, B. and Hjältén, J. (2000). Truncated power laws: a tool for understanding aggregation patterns in animals? *Ecology Letters*, **3**, 90–94.

Smith, A. B. (1994). *Systematics and the fossil record: documenting evolutionary patterns.* Blackwell Science, Oxford.

Smith, A. B. and Jeffery, C. H. (1998). Selectivity of extinction among sea urchins at the end of the Cretaceous period. *Nature*, **392**, 69–71.

Smith, C. R. (ed.) (1990). *Handbook for atlasing American breeding birds.* Vermont Institute for Natural Science, Woodstock.

Smith, F. D. M., May, R. M. and Harvey, P. H. (1994). Geographical ranges of Australian mammals. *Journal of Animal Ecology*, **63**, 441–450.

Smith, K. W., Dee, C. W., Fearnside, J. D., Fletcher, E. W. and Smith, R. N. (eds) (1993). *The breeding birds of Hertfordshire.* Hertfordshire Natural History Society.

Smith, L. (1992). Effect of temperature on life history characteristics of *Anisopteromalus calandrae* (Hymenoptera: Pteromalidae) parasitizing maize weevil larvae in corn kernels. *Environmental Entomology*, **21**, 877–887.

Smith, R. B. (1972). Relation of topography and vegetation to the occurrence of douglas-fir dwarf mistletoe at its northern limits in British Columbia. *Ecology*, **53**, 729–734.

Smith, T. B., Wayne, R. K., Girman, D. J. and Bruford, M. W. (1997). A role for ecotones in generating rainforest biodiversity. *Science*, **276**, 1855–1857.

Snyder, D. J. and Peterson, M. S. (1999). Life history of a peripheral population of bluespotted sunfish *Enneacanthus gloriosus* (Holbrook), with comments on geographic variation. *American Midland Naturalist*, **141**, 345–357.

Soberón, J. and Koleff, P. (1997). The national biodiversity information system of Mexico. In *Nature and human society: the quest for a sustainable world* (ed. P. H. Raven), pp. 586–595. National Academy Press, Washington, D. C.

Soberón, M. J. and Loevinsohn, M. (1987). Patterns of variations in the numbers of animal populations and the biological foundations of Taylor's law of the mean. *Oikos*, **48**, 249–252.

Solem, A. (1984). A world model of land snail diversity and abundance. In *World-wide snails: biogeographical studies on non-marine Mollusca* (eds A. Solem and A. C. van Bruggen), pp. 6–22. E. J. Brill/W. Backhuys, Leiden.

Soler, M. and Soler, J. J. (1992). Latitudinal trends in clutch size in single brooded hole nesting bird species: a new hypothesis. *Ardea*, **80**, 293–300.

Soltis, D. E., Gitzendanner, M. A., Strenge, D. D. and Soltis, P. S. (1997). Chloroplast DNA intraspecific phylogeography of plants from the Pacific Northwest of North America. *Plant Systematics and Evolution*, **206**, 353–373.

Soltz, D. L. and Naiman, R. J. (1978). *The natural history of native fishes in the Death Valley system.* Natural History Museum of Los Angeles County, Los Angeles.

Soulé, M. (1973). The epistasis cycle: a theory of marginal populations. *Annual Review of Ecology and Systematics*, **4**, 165–187.

Soulé, M. E. and Sanjayan, M. A. (1988). Ecology–conservation targets: do they help? *Science*, **279**, 2060–2061.

Southward, A. J. and Crisp, D. J. (1954). Recent changes in the distribution of the intertidal barnacles *Chthamalus stellatus* Poli and *Balanus balanoides* L. in the British Isles. *Journal of Animal Ecology*, **23**, 163–177.

Spencer, M. (2000). Are predators rare? *Oikos*, **89**, 115–122.

Spicer, J. I. and Gaston, K. J. (1999). *Physiological diversity and its ecological implications.* Blackwell Science, Oxford.

Spradbery, J. P. and Maywald, G. F. (1992). The distribution of the european or german wasp, *Vespula germanica* (F.) (Hymenoptera: Vespidae), in Australia: past, present and future. *Australian Journal of Zoology*, **40**, 495–510.

St Clair, R. C. and Gregory, P. T. (1990). Factors affecting the northern range limit of painted turtles (*Chrysemys picta*): winter acidosis or freezing? *Copeia*, 1990, 1083–1089.

Standing, K. L., Herman, T. B., Hurlburt, D. D. and Morrison, I. B. (1997). Post-emergence behaviour of neonates in a northern peripheral population of Blanding's turtle, *Emydoidea blandingii*, in Nova Scotia. *Canadian Journal of Zoology*, **75**, 1387–1395.

Standley, P., Bucknell, N. J., Swash, A. and Collins, I. D. (1996). *The birds of Berkshire.* Berkshire Atlas Group, Reading.

Stanley, S. M. (1979). *Macroevolution: patterns and process.* W. H. Freeman, San Francisco.

Stanley, S. M. (1986). Population size, extinction, and speciation: the fission effect in Neogene Bivalvia. *Paleobiology*, **12**, 89–110.

Stanley, S. M., Wetmore, K. L. and Kennett, J. P. (1988). Macroevolutionary differences between the two major clades of Neogene planktonic foraminifera. *Paleobiology*, **14**, 235–249.

Stattersfield, A. J., Crosby, M. J., Long, A. J. and Wege, D. C. (1998). *Endemic bird areas of the world: priorities for biodiversity conservation.* BirdLife International, Cambridge.

Steadman, D. W. (1995). Prehistoric extinctions of Pacific Island birds: biodiversity meets zooarchaeology. *Science*, **267**, 1123–1131.

Stebbins, G. L. (1978). Why are there so many rare plants in California? II. Youth and age of species. *Fremontia*, **6**, 17–20.

Stebbins, G. L. and Major, J. (1965). Endemism and speciation in the California flora. *Ecological Monographs*, **35**, 1–35.

Stephen, A. C. (1938). Temperature and the incidence of certain species in western european waters in 1932–1934. *Journal of Animal Ecology*, **7**, 125–129.

Stevens, G. C. (1989). The latitudinal gradient in geographical range: how so many species co-exist in the tropics. *American Naturalist*, **133**, 240–256.

Stevens, G. (1992). Spilling over the competitive limits to species coexistence. In *Systematics, ecology, and the biodiversity crisis* (ed. N. Eldredge), pp. 40–58. Columbia University Press, New York.

Stevens, G. C. (1996). Extending Rapoport's rule to Pacific marine fishes. *Journal of Biogeography*, **23**, 149–154.

Stevens, G. C. and Enquist, B. J. (1997). Macroecological limits to the abundance and distribution of *Pinus*. In *Ecology and biogeography of Pinus* (ed. D. M. Richardson), pp. 183–190. Cambridge University Press, Cambridge.

Stewart-Oaten, A., Murdoch, W. W. and Walde, S. J. (1995). Estimation of temporal variability in populations. *American Naturalist*, **146**, 519–535.

Stöcklin, J. and Körner, C. (1999). Recruitment and mortality of *Pinus sylvestris* near the Nordic treeline: the role of climatic change and herbivory. *Ecological Bulletins*, **47**, 168–177.

Stork, N. E. (1997). Measuring global biodiversity and its decline. In *Biodiversity II: understanding and protecting our biological resources* (eds M. L. Reaka-Kudla, D. E. Wilson and E. O. Wilson), pp. 41–68. Joseph Henry Press, Washington, D. C.

Stotz, D. F., Fitzpatrick, J. W., Parker, T. A. and Moskovits, D. K. (1996). *Neotropical birds. Ecology and conservation.* University of Chicago Press, Chicago.

Strathdee, A. T. and Bale, J. S. (1995). Factors limiting the distribution of *Acyrthosiphon svalbardicum* (Hemiptera: Aphididae) on Spitsbergen. *Polar Biology*, **15**, 375–380.

Straw, N. A. (1994). Species-area relationships and population dynamics: two sides of the same coin. In *Individuals, populations and patterns in ecology* (eds S. R. Leather, A. D. Watt, N. J. Mills and K. F. A. Walters), pp. 275–286. Intercept, Andover.

Strayer, D. L. (1991). Projected distribution of the zebra mussel, *Dreissena polymorpha*, in North America. *Canadian Journal of Fisheries and Aquatic Sciences*, **48**, 1389–1395.

Strayer, D. L. (2001). Endangered freshwater invertebrates. In *Encyclopedia of biodiversity*, *Vol. 2* (ed. S. A. Levin), pp. 425–439. Academic Press, San Diego.

Strong, D. R., Lawton, J. H. and Southwood, T. R. E. (1984). *Insects on plants: community patterns and mechanisms*. Blackwell Scientific, Oxford.

Sutherland, W. J. and Baillie, S. R. (1993). Patterns in the distribution, abundance and variation of bird populations. *Ibis*, **135**, 209–210.

Svensson, B. W. (1992). Changes in occupancy, niche breadth and abundance of three *Gyrinus* species as their respective range limits are approached. *Oikos*, **63**, 147–156.

Swain, D. P. and Morin, R. (1996). Relationships between geographic distribution and abundance of American plaice (*Hippoglossoides platessoides*) in the southern Gulf of St. Lawrence. *Canadian Journal of Fisheries and Aquatic Sciences*, **53**, 106–119.

Swain, D. P. and Sinclair, A. F. (1994). Fish distribution and catchability: what is the appropriate measure of distribution? *Canadian Journal of Fisheries and Aquatic Sciences*, **51**, 1046–1054.

Swain, D. P. and Wade, E. J. (1993). Density-dependent geographic distribution of Atlantic cod (*Gadus morhua*) in the southern Gulf of St. Lawrence. *Canadian Journal of Fisheries and Aquatic Sciences*, **50**, 725–733.

Swihart, R. K., Slade, N. A. and Bergstrom, B. J. (1988). Relating body size to the rate of home range use in mammals. *Ecology*, **69**, 393–399.

Sykes, M. T., Prentice, I. C. and Cramer, W. (1996). A bioclimatic model for the potential distributions of north European tree species under present and future climates. *Journal of Biogeography*, **23**, 203–233.

Tanaka, S. and Suzuki, Y. (1998). Physiological trade-offs between reproduction, flight capability and longevity in a wing-dimorphic cricket, *Modicogryllus confirmatus*. *Journal of Insect Physiology*, **44**, 121–129.

Tansley, S. A. (1988). The status of threatened Proteaceae in the Cape Flora, South Africa, and the implications for their conservation. *Biological Conservation*, **43**, 227–239.

Taper, M. L., Böhning-Gaese, K. and Brown, J. H. (1995). Individualistic responses of bird species to environmental change. *Oecologia*, **101**, 478–486.

Tauber, C. and Tauber, M. J. (1989). Sympatric speciation in insects: perception and perspective. In *Speciation and its consequences* (eds D. Otte and J. A. Endler), pp. 307–344. Sinauer Associates, Sunderland, Massachusetts.

Taulman, J. F. and Robbins, L. W. (1996). Recent range expansion and distributional limits of the none-banded armadillo (*Dasypus novemcinctus*) in the United States. *Journal of Biogeography*, **23**, 635–648.

Taylor, B. and van Perlo, B. (1998). *Rails: a guide to the rails, crakes, gallinules and coots of the World*. Pica Press, Mountfield.

Taylor, C. M. and Gotelli, N. J. (1994). The macroecology of *Cyprinella*: correlates of phylogeny, body size, and geographical range. *American Naturalist*, **144**, 549–569.

Taylor, L. R. (1961). Aggregation, variance and the mean. *Nature*, **189**, 732–735.

Taylor, L. R. (1977). Migration and the spatial dynamics of an aphid, *Myzus persicae*. *Journal of Animal Ecology*, **46**, 411–423.

Taylor, L. R. (1984). Assessing and interpreting the spatial distributions of insect populations. *Annual Review of Entomology*, **29**, 321–357.

Taylor, L. R. (1986). Synoptic dynamics, migration and the Rothamsted insect survey. *Journal of Animal Ecology*, **55**, 1–38.

Taylor, L. R. and Woiwod, I. P. (1982). Comparative synoptic dynamics. I. Relationships between inter- and intraspecific spatial and temporal variance/mean parameters. *Journal of Animal Ecology*, **51**, 879–906.

Taylor, L. R., Woiwod, I. P. and Perry, J. N. (1978). The density-dependence of spatial behaviour and the rarity of randomness. *Journal of Animal Ecology*, **47**, 383–406.

Taylor, L. R., Woiwod, I. P. and Perry, J. N. (1979). The negative binomial as a dynamic ecological model for aggregation and the density dependence of *k*. *Journal of Animal Ecology*, **48**, 289–304.

Taylor, L. R., Woiwod, I. P. and Perry, J. N. (1980). Variance and the large scale spatial stability of aphids, moths and birds. *Journal of Animal Ecology*, **49**, 831–854.

Taylor, R. A. J. and Taylor, L. R. (1979). A behavioural model for the evolution of spatial dynamics. In *Population dynamics* (eds R. M. Anderson, B. D. Turner and L. R. Taylor), pp. 1–27. Blackwell Science, Oxford.

Tellería, J. L. and Santos, T. (1993). Distributional patterns of insectivorous passerines in the Iberian forests: does abundance decrease near the border? *Journal of Biogeography*, **20**, 235–240.

Tellería, J. L. and Santos, T. (1999). Distribution of birds in fragments of Mediterranean forests: the role of ecological densities. *Ecography*, **22**, 13–19.

Terborgh, J. (1973). On the notion of favorableness in plant ecology. *American Naturalist*, **107**, 481–501.

Terborgh, J. (1985). The role of ecotones in the distribution of Andean birds. *Ecology*, **66**, 1237–1246.

Terborgh, J. and Winter, B. (1980). Some causes of extinction. In *Conservation biology: an evolutionary-ecological perspective* (eds M. E. Soulé and B. A. Wilcox), pp. 119–133. Sinauer Associates, Sunderland, Massachusetts.

Terry, A., Bucciarelli, G. and Bernardi, G. (2000). Restricted gene flow and incipient speciation in disjunct Pacific Ocean and Sea of Cortez populations of a reef fish species, *Girella nigricans*. *Evolution*, **54**, 652–659.

Thaung, M. and Collins, P. J. (1986). Joint effects of temperature and insecticides on mortality and fecundity of *Sitophilus oryzae* (Coleoptera: Curculionidae) in wheat and maize. *Journal of Economic Entomology*, **79**, 909–914.

Thomas, C. D. (2000). Dispersal and extinction in fragmented landscapes. *Proceedings of the Royal Society, London B*, **267**, 139–145.

Thomas, C. D. and Abery, J. C. G. (1995). Estimating rates of butterfly decline from distribution maps: the effect of scale. *Biological Conservation*, **73**, 59–65.

Thomas, C. D., Jordano, D., Lewis, O. T., Hill, J. K., Sutcliffe, O. L. and Thomas, J. A. (1998). Butterfly distributional patterns, processes and conservation. In *Conservation in a changing world* (eds G. M. Mace, A. Balmford and J. R. Ginsberg), pp. 107–138. Cambridge University Press, Cambridge.

Thomas, C. D. and Lennon, J. J. (1999). Birds extend their ranges northwards. *Nature*, **399**, 213.

Thomas, C. D., Thomas, J. A. and Warren, M. S. (1992). Distributions of occupied and vacant butterfly habitats in fragmented landscapes. *Oecologia*, **92**, 563–567.

Thomas, J. A. (1993). Holocene climate changes and warm man-made refugia may explain why a sixth of British butterflies possess unnatural early-successional habitats. *Ecography*, **16**, 278–284.

Thomas, J. A. (1994). Why small cold-blooded insects pose different conservation problems to birds in modern landscapes. *Ibis*, **137**, S112–S119.

Thomas, J. A., Moss, D. and Pollard, E. (1994). Increased fluctuations of butterfly populations towards the northern edges of species' ranges. *Ecography*, **17**, 215–220.

Thompson, K. Hodgson, J. G. and Gaston, K. J. (1998). Abundance-range size relationships in the herbaceous flora of central England. *Journal of Ecology*, **86**, 439–448.

Thomson, J. D., Weiblen, G., Thomson, B. A., Alfaro, S. and Legendre, P. (1996). Untangling multiple factors in spatial distributions: lilies, gophers, and rocks. *Ecology*, **77**, 1698–1715.

Thorson, G. (1950). Reproductive and larval ecology of marine bottom invertebrates. *Biological Reviews*, **25**, 1–45.

Tiainen, J., Hanski, I. P. and Mehtälä, J. (1983). Insulation of nests and the northern limits of three *Phylloscopus* warblers in Finland. *Ornis Scandinavica*, **14**, 149–153.

Tigerstedt, P. M. A. (1973). Studies on isozyme variation in marginal and central populations of *Picea abies*. *Hereditas*, **75**, 47–60.

Tilman, D., Lehman, C. L. and Kareiva, P. (1997). Population dynamics in spatial habitats. In *Spatial ecology* (eds D. Tilman and P. Kareiva), pp. 3–20. Princeton University Press, New Jersey.

Tokeshi, M. (1996). Power fraction: a new explanation of relative abundance patterns in species-rich assemblages. *Oikos*, **75**, 543–550.

Tompkins, D. M., Greenman, J. V., Robertson, P. A. and Hudson, P. J. (2000). The role of shared parasites in the exclusion of wildlife hosts: *Heterakis gallinarum* in the ring-necked pheasant and the grey partridge. *Journal of Animal Ecology*, **69**, 829–840.

Towns, D. R. and Daugherty, C. H. (1994). Patterns of range contractions and extinctions in the New Zealand herpetofauna following human colonisation. *New Zealand Journal of Zoology*, **21**, 325–339.

Tucker, G. M. and Heath, M. F. (1994). *Birds in Europe: their conservation status*. BirdLife International, Cambridge.

Turner, A. and Antón, M. (1997). *The big cats and their fossil relatives*. Columbia University Press, New York.

Twomey, A. C. (1936). Climographic studies of certain introduced and migratory birds. *Ecology*, **17**, 122–132.

Tyler, J. A. and Hargrove, W. W. (1997). Predicting spatial distribution of foragers over large resource landscapes: a modeling analysis of the Ideal Free Distribution. *Oikos*, **79**, 376–386.

Udvardy, M. D. F. (1969). *Dynamic zoogeography: with special reference to land animals*. Van Nostrand Reinhold, New York.

Ungerer, M. J., Ayres, M. P. and Lombardero, M. J. (1999). Climate and the northern distribution limits of *Dendroctonus frontalis* Zimmermann (Coleoptera: Scolytidae). *Journal of Biogeography*, **26**, 1133–1145.

Vander Haegen, W. M., Dobler, F. C. and Pierce, D. J. (2000). Shrubsteppe bird response to habitat and landscape variables in eastern Washington, U. S. A. *Conservation Biology*, **14**, 1145–1160.

van Herrewege, J. and David, J. R. (1997). Starvation and desiccation tolerances in *Drosophila*: Comparison of species from different climatic origins. *Écoscience*, **4**, 151–157.

Van Horne, B. (1983). Density as a misleading indicator of habitat quality. *Journal of Wildlife Management*, **47**, 893–901.

Van Riel, P., Jordaens, K., Van Goethem, J. L. and Backeljau, T. (2001). Genetic variation in the land snail *Isognomostoma isognomostoma* (Gastropoda: Pulmonata: Helicidae). *Malacologia*, **45**, 1–11.

van Rossum, F., Vekemans, X., Meerts, P., Gratia, E. and Lefèbvre, C. (1997). Allozyme variation in relation to ecotypic differentiation and population size in marginal populations of *Silene nutans*. *Heredity*, **78**, 552–560.

van Swaay, C. A. M. (1995). Measuring changes in butterfly abundance in The Netherlands. In *Ecology and conservation of butterflies* (ed. A. S. Pullin), pp. 230–247. Chapman and Hall, London.

Van Valen, L. (1973a). A new evolutionary law. *Evolutionary Theory*, **1**, 1–30.

Van Valen, L. (1973b). Body size and numbers of plants and animals. *Evolution*, **27**, 27–35.

Veistola, S., Lehikoinen, E. and Isolivari, L. (1995). Breeding biology of the Great Tit *Parus major* in a marginal population in northernmost Finland. *Ardea*, **83**, 419–420.

Veit, R. R. (2000). Vagrants as the expanding fringe of a growing population. *Auk*, **117**, 242–246.

Veit, R. R. and Lewis, M. A. (1996). Dispersal, population growth, and the Allee effect: dynamics of the house finch invasion of eastern North America. *American Naturalist*, **148**, 255–274.

Venier, L. A. and Fahrig, L. (1998). Intra-specific abundance-distribution relationships. *Oikos*, **82**, 483–490.

Venier, L. A., McKenney, D. W., Wang, Y. and McKee, J. (1999). Models of large-scale breeding-bird distribution as a function of macro-climate in Ontario, Canada. *Journal of Biogeography*, **26**, 315–328.

Villard, M-A. and Maurer, B. A. (1996). Geostatistics as a tool for examining hypothesized declines in migratory songbirds. *Ecology*, **77**, 59–68.

Virgós, E. and Casanovas, J. G. (1999). Environmental constraints at the edge of a species distribution, the Eurasian badger (*Meles meles* L.): a biogeographic approach. *Journal of Biogeography*, **26**, 559–564.

Virkkala, R. (1993). Ranges of northern forest passerines: a fractal analysis. *Oikos*, **67**, 218–226.

Vitousek, P. M., Ehrlich, P. R., Ehrlich, A. H. and Matson, P. A. (1986). Human appropriation of the products of photosynthesis. *BioScience*, **36**, 368–373.

Voous, K. H. (1960). *Atlas of European birds*. Nelson, Edinburgh.

Vrba, E. (1993). Turnover-pulses, the red queen, and related topics. *American Journal of Science*, **293A**, 418–452.

Wagner, P. J. and Erwin, D. H. (1995). Phylogenetic patterns as tests of speciation models. In *New approaches to speciation in the fossil record* (eds D. H. Erwin and R. L. Anstey), pp. 87–122. Columbia University Press, New York.

Walck, J. L., Baskin, J. M. and Baskin, C. C. (2001). Why is *Solidago shortii* narrowly endemic and *S. altissima* geographically widespread? A comprehensive comparative study of biological traits. *Journal of Biogeography*, **28**, 1221–1237.

Wallace, A. R. (1849). On the monkeys of the Amazon. *Proceedings of the Zoological Society, London*, **20**, 107–110.

Ward, S. A., Sunderland, K. D., Chambers, R. J. and Dixon, A. F. G. (1986). The use of incidence counts for estimation of cereal aphid populations. 3. Population development and the incidence–density relation. *Netherlands Journal of Plant Pathology*, **92**, 175–183.

Warren, P. H. and Gaston, K. J. (1997). Interspecific abundance-occupancy relationships: a test of mechanisms using microcosms. *Journal of Animal Ecology*, **66**, 730–742.

Watkinson, A. R. (1985). On the abundance of plants along an environmental gradient. *Journal of Ecology*, **73**, 569–578.

Watkinson, A. R., Freckleton, R. P. and Forrester, L. (2000). Population dynamics of *Vulpia ciliata*: regional, patch and local dynamics. *Journal of Ecology*, **88**, 1012–1029.

Watkinson, A. R. and Sutherland, W. J. (1995). Sources, sinks and pseudo-sinks. *Journal of Animal Ecology*, **64**, 126–130.

Watson, L. (1981). *Whales of the world*. Hutchinson, London.

Watts, P. C., Thorpe, J. P. and Taylor, P. D. (1998). Natural and anthropogenic dispersal mechanisms in the marine environment: a study using cheilostome Bryozoa. *Philosophical Transactions of the Royal Society, London B*, **353**, 453–464.

WCMC (World Conservation Monitoring Centre) (1992). *Global biodiversity: status of the Earth's living resources*. Chapman and Hall, London.

Webb, S. L., Glenn, M. G., Cook, E. R., Wagner, W. S. and Thetford, R. D. (1993). Range edge red spruce in New Jersey, U. S. A.: bog versus upland population structure and climate responses. *Journal of Biogeography*, **20**, 63–78.

Webb, T. J. and Gaston, K. J. (2000). Geographic range size and evolutionary age in birds. *Proceedings of the Royal Society, London B*, **267**, 1843–1850.

Webb, T. J., Kershaw, M. and Gaston, K. J. (2001). Rarity and phylogeny in birds. In *Biotic homogenization: the loss of diversity through invasion and extinction* (eds J. L. Lockwood and M. L. McKinney), pp. 57–80. Kluwer Academic/Plenum Publishers, New York.

Weber, E. (2001). Current and potential ranges of three exotic goldenrods (*Solidago*) in Europe. *Conservation Biology*, **15**, 122–128.

Weber, E. F. (1997). The alien flora of Europe: a taxonomic and biogeographic review. *Journal of Vegetation Science*, **8**, 565–572.

Weidema, I., Siegismund, H. R. and Philipp, M. (1996). Distribution of genetic variation within and among Danish populations of *Armeria maritima*, with special reference to the effects of population size. *Hereditas*, **124**, 121–129.

Wendel, J. F. and Parks, C. R. (1985). Genetic diversity and population structure in *Camellia japonica* L. (Theaceae). *American Journal of Botany*, **72**, 52–65.

Westman, W. E. (1980). Gaussian analysis: identifying environmental factors influencing bell-shaped species distributions. *Ecology*, **61**, 733–739.

Westman, W. E. (1991). Measuring realized niche spaces: climatic response of chaparral and coastal sage scrub. *Ecology*, **72**, 1678–1684.

Westrop, S. R. (1989). Macroevolutionary implications of mass extinction—evidence from an Upper Cambrian stage boundary. *Paleobiology*, **15**, 46–52.

Westrop, S. R. (1991). Intercontinental variation in mass extinction patterns: influence of biogeographic province. *Paleobiology*, **17**, 363–368.

Westrop, S. R. and Lundvigsen, R. (1987). Biogeographic control of trilobite mass extinction at an Upper Cambrian 'biomere' boundary. *Paleobiology*, **13**, 84–99.

Whitcomb, R. F., Hicks, A. L., Blocker, H. D. and Lynn, D. E. (1994). Biogeography of leafhopper specialists of the shortgrass prairie: evidence for the roles of phenology in determination of biological diversity. *American Entomologist*, Spring, 19–35.

Whittaker, J. B. (1971). Population changes in *Neophilaenus lineatus* (L.) (Homoptera: Cercopidae) in different parts of its range. *Journal of Animal Ecology*, **40**, 425–443.

Whittaker, J. B. and Tribe, N. P. (1996). An altitudinal transect as an indicator of responses of a spittlebug (Auchenorrhyncha: Cercopidae) to climate change. *European Journal of Entomology*, **93**, 319–324.

Wiencke, C., Bartsch, I., Bischoff, B., Peters, A. F. and Breeman, A. M. (1994). Temperature requirements and biogeography of Antarctic, Arctic and Amphiequatorial seaweeds. *Botanica Marina*, **37**, 247–259.

Wiens, J. A. (1989). *The ecology of bird communities*, Vol. 1. Foundations and patterns. Cambridge University Press, Cambridge.

Wilcove, D. S. (2000). *The condor's shadow: the loss and recovery of wildlife in America*. Anchor Books, New York.

Wilcove, D. S. and Terborgh, J. W. (1984). Patterns of population decline in birds. *American Birds*, **38**, 10–13.

Wilkinson, D. M. (2001). What is the upper size limit for cosmopolitan distribution in free-living microorganisms? *Journal of Biogeography*, **28**, 285–291.

Williams, C. B. (1964). *Patterns in the balance of nature*. Academic Press, London.

Williams, P. H. (1988). Habitat use by bumblebees (*Bombus* spp.). *Ecological Entomology*, **13**, 223–237.

Williams, P. H. (1991). The bumble bees of the Kashmir Himalaya (Hymenoptera: Apidae, Bombini). *Bulletin of the British Museum Natural History (Entomology)*, **60**, 1–204.

Williams, P. H. (1998). An annotated checklist of bumble bees with an analysis of patterns of description (Hymenoptera: Apidae, Bombini). *Bulletin of The Natural History Museum (Entomology)*, **67**, 79–152.

Williams, P. H. and Araújo, M. B. (2000). Using probability of persistence to identify important areas for biodiversity conservation. *Proceedings of the Royal Society, London B*, **267**, 1959–1966.

Williams, P. H., Burgess, N. and Rahbek, C. (2000). Assessing large 'flagship species' for representing the diversity of sub-Saharan mammals. In *Priorities for the conservation of mammalian diversity* (eds A. Entwistle and N. Dunstone), pp. 85–99. Cambridge University Press, Cambridge.

Williamson, M. H. (1957). An elementary theory of interspecific competition. *Nature*, **180**, 422–425.

Williamson, M. H. (1972). *The analysis of biological populations*. Arnold, London.

Williamson, M. (1996). *Biological invasions*. Chapman and Hall, London.

Williamson, M. and Gaston, K. J. (1999). A simple transformation for sets of range sizes. *Ecography*, **22**, 674–680.

Williamson, M. H. and Lawton, J. H. (1991). Fractal geometry of ecological habitats. In *Habitat structure: the physical arrangement of objects in space* (eds S. S. Bell, E. D. McCoy and H. R. Mushinsky), pp. 69–86. Chapman and Hall, London.

Willis, A. J. (1985). Plant diversity and change in a species-rich dune system. *Transactions of the Botanical Society of Edinburgh*, **44**, 291–308.

Willis, J. C. (1922). *Age and area: a study in geographical distribution and origin of species*. Cambridge University Press, Cambridge.

Wilson, E. O. (1961). The nature of the taxon cycle in the Melanesian ant fauna. *American Naturalist*, **95**, 169–193.

Wilson, E. O. (1992). *The diversity of life*. Penguin, London.

Wilson, E. O. and Peter, F. M. (eds) (1988). *BioDiversity*. National Academy Press, Washington, D. C.

Wilson, J. B., Ronghua, Y., Mark, A. F. and Agnew, A. D. Q. (1991). A test of the low marginal variance (LMV) theory, in *Leptospermum scoparium* (Myrtaceae). *Evolution*, **45**, 780–784.

Wilson, S. D. and Keddy, P. A. (1988). Species richness, survivorship, and biomass accumulation along an environmental gradient. *Oikos*, **53**, 375–380.

Wilson, W. G., Nisbet, R. M., Ross, A. H., Robles, C. and Desharnais, R. A. (1996). Abrupt population changes along smooth environmental gradients. *Bulletin of Mathematical Biology*, **58**, 907–922.

Wing, L. (1943). Spread of the starling and English sparrow. *Auk*, **60**, 74–87.

Winston, M. R. and Angermeier, P. L. (1995). Assessing conservation value using centers of population density. *Conservation Biology*, **9**, 1518–1527.

Winters, G. H. and Wheeler, J. P. (1985). Interaction between stock area, stock abundance, and catchability coefficient. *Canadian Journal of Fisheries and Aquatic Sciences*, **42**, 989–998.

Wolf, A. T., Harrison, S. P. and Hamrick, J. L. (2000). Influence of habitat patchiness on genetic diversity and spatial structure of a serpentine endemic plant. *Conservation Biology*, **14**, 454–463.

Wolf, C. M., Garland, T. Jr and Griffith, B. (1998). Predictors of avian and mammalian translocation success: reanalysis with phylogenetically independent contrasts. *Biological Conservation*, **86**, 243–255.

Wolf, P. G., Schneider, H. and Ranker, T. A. (2001). Geographic distributions of homosporous ferns: does dispersal obscure evidence of vicariance? *Journal of Biogeography*, **28**, 263–270.

Wood, G. L. (1982). *The Guinness book of animal facts and feats*. Guinness Superlatives, Enfield, Middlesex.

Woodroffe, R. (2000). Predators and people: using human densities to interpret declines of large carnivores. *Animal Conservation*, **3**, 165–173.

Woods, K. D. and Davis, M. B. (1989). Paleoecology of range limits: beech in the upper peninsula of Michigan. *Ecology*, **70**, 681–696.

Woodward, F. I. (1987). *Climate and plant distribution*. Cambridge University Press, Cambridge.

Woodward, F. I. (1990). The impact of low temperatures in controlling the geographical distribution of plants. *Philosophical Transactions of the Royal Society, London B*, **326**, 585–593.

Woodward, F. I. (1997). Life at the edge: a 14-year study of a *Verbena officinalis* population's interactions with climate. *Journal of Ecology*, **85**, 899–906.

Woodward, F. I. and Pigott, C. D. (1975). The climatic control of the altitudinal distribution of *Sedum rosea* (L.) Scop. and *S. telephium* L. I. Field observations. *New Phytologist*, **74**, 323–334.

Wright, D. H. (1991). Correlations between incidence and abundance are expected by chance. *Journal of Biogeography*, **18**, 463–466.

Yagami, T. and Goto, A. (2000). Patchy distribution of a fluvial sculpin, *Cottus nozawae*, in the Gakko River system at the southern margin of its native range. *Ichthyological Research*, **47**, 277–286.

Yamashita, T. and Polis, G. A. (1995). A test of the central-marginal model using sand scorpion populations (*Paruroctonus mesaensis*, Vaejovidae). *Journal of Arachnology*, **23**, 60–64.

Yaninek, J. S., Gutierrez, A. P. and Herren, H. R. (1989). Dynamics of *Mononychellus tanajoa* (Acari: Tetranychidae) in Africa: experimental evidence of temperature and host plant effects on population growth rates. *Environmental Entomology*, **18**, 633–640.

Yeatman, L. (1976). *Atlas des oiseaux nicheurs de France 1970 à 1975.* Société Ornithologique de France, Paris.

Yeatman-Berthelot, D. and Jarry, G. (1994). *Nouvel atlas des oiseaux nicheurs de France 1985–1989.* Société Ornithologique de France, Paris.

Yeh, F. C. and Layton, C. (1979). The organization of genetic variability in central and marginal populations of lodgepole pine *Pinus contorta* spp. *latifolia. Canadian Journal of Genetics and Cytology*, **21**, 487–503.

Yeh, F. C. and O'Malley, D. (1980). Enzyme variations in natural populations of Douglas fir (*Pseudotsuga menziesii* (Mirb.) Franco) from British Columbia. I. Genetic variation patterns in coastal populations. *Silvae Genetica*, **29**, 93–92.

Yom-Tov, Y. (1979). Is air temperature limiting northern breeding distribution of birds? *Ornis Scandinavica*, **11**, 71–72.

Yom-tov, Y., Christie, M. I. and Iglesias, G. J. (1994). Clutch size in passerines of southern South America. *Condor*, **96**, 170–177.

Young, B. E. (1994). Geographic and seasonal patterns of clutch-size variation in house wrens. *Auk*, **111**, 545–555.

Zeisset, I. and Beebee, T. J. C. (2001). Determination of biogeographical range: an application of molecular phylogeography to the European pool frog *Rana lessonae. Proceedings of the Royal Society London*, B **268**, 933–938.

Zera, A. J. and Brink, T. (2000). Nutrient absorption and utilization by wing and flight muscle morphs of the cricket *Gryllus firmus*: implications for the trade-off between flight capability and early reproduction. *Journal of Insect Physiology*, **46**, 1207–1218.

Zeveloff, S. I. and Boyce, M. S. (1988). Body size patterns in North American mammal faunas. In *Evolution of life histories of mammals* (ed. M. S. Boyce), pp. 123–146. Yale University Press, New Haven.

Zheng, X. Y. and Eltahir, E. A. B. (1998). The role of vegetation in the dynamics of West African monsoons. *Journal of Climate*, **11**, 2078–2096.

Zwölfer, H. (1987). Species richness, species packing, and evolution in insect-plant systems. In *Ecological studies*, Vol. 61 (eds. E.-D. Schulze and H. Zwölfer), pp. 301–319. Springer-Verlag, Berlin.

Index

Lightning Source UK Ltd.
Milton Keynes UK
UKOW04f2023150814

236994UK00017B/104/P